高等院校"十三五"重点规划 专业英语系列

空管与签派专业英语阅读

Air Traffic Controller and Dispatcher

Professional English Reading

沈志远　主编

哈尔滨工程大学出版社
Harbin Engineering University Press

内 容 简 介

《空管与签派专业英语阅读》是一本供航空航天类高等院校的大学生学习民航专业英语的教材,特别面向空中交通管制和签派专业的学生。本书的题材涵盖了与民航业务相关的基本内容,不仅包含飞行史、飞机的结构、导航及机载设备、航空气象、飞行装载与配平、飞行程序设计等内容,也包含国家空域系统、机场协同决策、ADS - B 等一批最新的民航技术内容。本书的内容主要来源于近年来国外规章、杂志报道、学术期刊论文以及运行手册等。本书亦可作为民航系统各下属单位的飞行员、空乘人员、基地维修保养人员、行政管理人员学习和提高民航专业英语的教材和读物,以及广大民航爱好者学习民航英语的自学材料。

图书在版编目(CIP)数据

空管与签派专业英语阅读 / 沈志远主编. —哈尔滨:
哈尔滨工程大学出版社,2019.6
ISBN 978 - 7 - 5661 - 2229 - 2

Ⅰ.①空… Ⅱ.①沈… Ⅲ.①空中交通管制 – 英语 –
阅读教学 – 高等学校 – 教材②民用航空 – 机场 – 业务 – 英
语 – 阅读教学 – 高等学校 – 教材 Ⅳ.①V355.1
②F560.81

中国版本图书馆 CIP 数据核字(2019)第 054242 号

空管与签派专业英语阅读
KONGGUAN YU QIANPAIZHUANYE YINGYU YUEDU

选题策划　马佳佳
责任编辑　张忠远　李　想
封面设计　刘长友

出版发行　哈尔滨工程大学出版社
社　　址　哈尔滨市南岗区南通大街 145 号
邮政编码　150001
发行电话　0451 - 82519328
传　　真　0451 - 82519699
经　　销　新华书店
印　　刷　哈尔滨市石桥印务有限公司
开　　本　787 mm×1 092 mm　1/16
印　　张　16.25
字　　数　470 千字
版　　次　2019 年 6 月第 1 版
印　　次　2019 年 6 月第 1 次印刷
定　　价　49.80 元
http://www.hrbeupress.com
E-mail:heupress@ hrbeu.edu.cn

前　言

PREFACE

As the most basic and important professional language in the civil aviation, English has always been one of the most important working technical for the subordinate units of civil aviation administration of China.

Air traffic control and dispatcher professional English reading is a textbook for college students in aviation and aerospace universities to learn professional English of civil aviation, especially for students majoring in air traffic control and airline dispatching. This book can also be used as a teaching material and reading material to learn and improve the professional English of civil aviation for various subordinate units of civil aviation system, such as pilots, flight attendants, maintenance staff and administrative staff, as well as a self-study material for civil aviation enthusiasts to learn civil aviation English.

This book covers the basic contents related to civil aviation. It includes not only flight history, aircraft structure, navigation and airborne equipment, aviation weather, flight loading and balancing, flight procedure design, but also a batch of the state-of-the-art civil aviation technologies and development directions such as national airspace system, airport collaborative decision-making are contained. The content and language of this book are novel, and the main information comes from the regulations, magazines and periodicals published abroad in recent years.

There are 18 units in the book. Each unit consists of 4 part, like Reading Text, Words and Expressions, Exercises and Supplement Reading.

Shen Zhiyuan was responsible for the topic for each unit, reading text and words list selection, and the compilation of exercises. Zhou Siyao and Chang Yujin provide a supplementation and modification for the words list, exercises and exercises answers.

The compilation and publication of this book has been concerned and guided by the leaders of civil aviation college of Nanjing University of Aeronautics and Astronautics. I would like to thank Nanjing University of Aeronautics and Astronautics for its "13th five-year plan" professional construction program for providing this book with funding.

Considering to the limited level of the author, the any suggestions and improvement from the readers are welcomed.

Shen Zhiyuan

Nanjing, 2019

前　言

PREFACE

英语作为民航业最基础和最重要的工作语言,一直以来是中国民用航空局下属各单位高度重视的学习任务之一。

《空管与签派专业英语阅读》是一本供航空航天类高等院校的大学生学习民航专业英语的教材,特别面向空中交通管制和航空公司签派专业的学生。本书亦可作为民航系统各下属单位的飞行员、空乘人员、基地维修保养人员、行政管理人员学习和提高民航专业英语的教材和读物,以及广大民航爱好者学习民航英语的自学材料。

本书的题材涵盖了与民航业务相关的基本内容,不仅包含飞行史、飞机的结构、导航及机载设备、航空气象、飞行装载与配平、飞行程序设计等内容,也包含对国家空域系统、机场协同决策等一批最新的民航技术和发展方向的介绍。本书的内容和语言新颖,主要资料来源于近年来国外出版的航空规章、期刊、论文等。

本书共 18 单元,每个单元课文由 Reading Text、Words and Expressions、Exercises 和 Supplement Reading 组成。

沈志远负责全书的构思、选材、生词表、练习的编写工作,周思遥和常雨劲对生词表、练习及答案做了补充和修改。

本书的编写和出版得到了南京航空航天大学民航学院有关领导的关心和指导,在此深表谢意。感谢南京航空航天大学"十三五"专业建设项目为本书出版提供的资助。

由于编者水平有限,书中欠妥和错误之处在所难免,敬请读者指正。

沈志远

2019 年于南京

目录 CONTENTS

Unit 1

History of Aviation

I   Reading Text

1   Primitive beginnings

The origin of mankind's desire to fly lost in the distant past. From the earliest legends there have been stories of men strapping birdlike wings, stiffened cloaks or other devices to themselves and attempting to fly, typically by jumping off a tower. The Greek legend of Daedalus and Icarus is one of the earliest known, others originated from India, China and the European Dark Ages. During this early period the issues of lift, stability and control were not understood, and most attempts ended in serious injury or death.

The kite may have been the first form of man-made aircraft[Figure 1]. It was invented in China possibly as far back as the 5th century BC by Mozi and Lu Ban. The later design often emulated flying insects, birds, and other beasts, both real and mythical. Kites spread from China around the world. After its introduction into India, the kite further evolved into the fighter kite, where an abrasive line is used to cut down other kites.

From ancient times the Chinese have understood that hot air rises and have applied the principle to a type of small hot air balloon called sky lantern. A sky lantern consists of a paper balloon under or just inside which a small lamp is placed. Sky lanterns are traditionally launched for pleasure during festivals. Their military application is attributed to the general Zhuge Liang, who is said to have used them to scare the enemy troops.

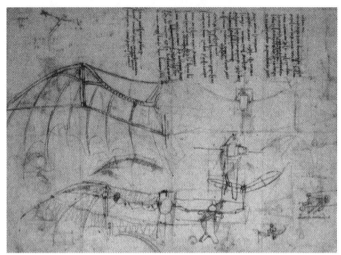

Figure 1　Man-made aircraft

2　The Renaissance

Eventually some investigators began to discover and define some basics of rational aircraft design. The most notable of these was Leonardo da Vinci, although his work remained unknown until 1797, and so had no influence on development over the next three hundred years. Leonardo studied bird flight, analyzing it and anticipating many principles of aerodynamics. He did at least understand that "An object offers as much resistance to the air as the air does to the object." Newton would not publish the Third law of motion until 1687.

Ballooning became a major "rage" in Europe in the late 18th century, providing the first detailed understanding for the relationship between the altitude and atmosphere. 1783 was a watershed year for ballooning and aviation, between June 4th and December 1st several aviation firsts were achieved in France.

Work on developing a steerable (or dirigible) balloon (now called an airship) continued sporadically throughout the 19th century. The first powered, controlled, sustained lighter-than-air flight is believed to have taken place in 1852 when Henri Giffard flew 15 miles (24 km) in France, with a steam engine driven craft.

3　Heavier-than-air flight

Sir George Cayley was first called the "father of the aero plane" in 1846. During the last years of the previous century he had begun the first rigorous study of the physics of flight and would later design the first modern heavier-than-air craft [Figure 2]. Among his many achievements, the most important contributions to aeronautics include:

(1) Clarifying our ideas and laying down the principles of heavier-than-air flight.

(2) Reaching a scientific understanding of the principles of bird flight.

(3) Conducting scientific aerodynamic experiment and demonstrating drag and streamlining, movement of the center of pressure, and the increase in lift from curving the wing surface.

(4) Defining the modern aero plane configuration and comprising a fixed wing, fuselage and tail assembly.

(5) Demonstrations of manned, gliding flight.

(6) Setting out the principles of power-to-weight ratio in sustaining flight.

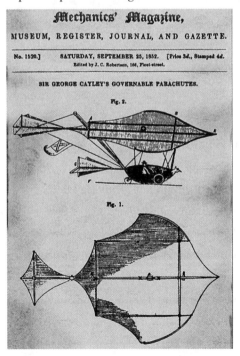

Figure 2　Sir George Cayley's Manuscript

The Wrights appear to be the first to make serious studied attempt to simultaneously solve the power and control problems. Using a methodological approach and concentrating on the controllability of the aircraft, the Wrights brothers had built and tested a series of kite and glider design from 1900 to 1902 before they attempted to build a powered design.

According to the Smithsonian Institution and Fédération Aéronautique Internationale (FAI), the Wrights made the first sustained, controlled, powered heavier-than-air manned flight at Kill Devil Hills, North Carolina, four miles (8 km) south of Kitty Hawk, North Carolina on December 17, 1903. The first flight by Orville Wright, of 120 feet (37 m) in 12 seconds, was recorded in a famous photograph[Figure 3].

Figure 3   First flight

4   World War I and World War II

It had seen great advancement in aircraft technology between World War I and World War II. Airplanes evolved from low-powered biplanes made from wood and fabric to sleek, high-powered monoplanes made of aluminum, primarily based on the founding work of Hugo Junkers during the World War I period and its adoption by American designer William Bushnell Stout and Soviet designer Andrei Tupolev. The age of the great rigid airships came and went. The first successful rotorcraft appeared in the form of the autogyro, invented by Spanish engineer Juan de la Cierva and first flown in 1919. In this design, the rotor is not powered but is spun like a windmill by its passage through the air. A separate powerplant is used to propel the aircraft forwards.

World War II achieved a great increase in the pace of development and production, of not only aircraft but also the associated flight-based weapon delivery systems. Air combat tactics and doctrines took advantage. Large-scale strategic bombing campaigns were launched, fighter escorts introduced and the more flexible aircraft and weapons allowed precise attacks on small targets with dive bombers, fighter-bombers, and ground-attack aircraft. New technologies like radar also allowed more coordinated and controlled deployment of air defense.

The first jet aircraft to fly was the Heinkel He 178 (Germany), flown by Erich Warsitz in 1939, followed by the world's first operational jet aircraft, the Me 262, in July 1942 and world's first jet-powered bomber, the Arado Ar 234, in June 1943.

5 The postwar era (1945—1979)

After World War II, commercial aviation grew rapidly, using mostly ex-military aircraft to transport people and cargo. This growth was accelerated by the glut of heavy and super-heavy bomber airframes like the B – 29 and Lancaster that could be converted into commercial aircraft.

USSR's Aeroflot became the first airline in the world to operate sustained regular jet services on September 15, 1956 with the Tupolev Tu – 104. The Boeing 707 and DC – 8 which established new level of comfort, safety and passenger expectation, ushered in the age of mass commercial air travel, dubbed the Jet Age.

The 1945 invention of nuclear bomb briefly increased the strategic importance of military aircraft in the Cold War between East and West. Even a moderate fleet of long-range bomber could deliver a deadly blow to the enemy, so great efforts were made to develop the countermeasure. At first, the supersonic interceptor aircraft were produced in considerable numbers. By 1955 most development efforts were shifted to guided surface-to-air missiles.

In 1979 the Gossamer Albatross became the first person to power aircraft to cross the English Channel. This achievement finally realized the dream of human flight for centuries.

6 The digital age

In 1986 Dick Rutan and Jeana Yeager flew an aircraft, the Rutan Voyager, around the world unrefueled, and without landing. In 1999 Bertrand Piccard became the first person to circle the earth in a balloon.

In the beginning of the 21st century, digital technology allowed subsonic military aviation to begin eliminating the pilot in favor of remotely operated or completely autonomous unmanned aerial vehicles(UAVs). In April 2001 the unmanned aircraft Global Hawk flew from Edwards AFB in the US to Australia non-stop and unrefueled. This was the longest point-to-point flight ever undertaken by an unmanned aircraft and took 23 hours and 23 minutes.

II   Words and Expressions

mankind	*n.*	人类;男性
strap	*vt.*	用带捆绑
stiffen	*vi.*	变硬;变猛烈;变黏
	vt.	使变硬;使黏稠
emulate	*vt.*	仿真;模仿;尽力赶上;同……竞争
evolve	*vt.*	发展;进化;使逐步形成;推断出
abrasive	*adj.*	粗糙的;有研磨作用的
	n.	研磨料
troop	*n.*	军队;组;群;多数
	vi.	群集;成群而行;结队
renaissance	*n.*	新生;再生;复活
investigator	*n.*	研究者;调查者;侦查员
rational	*adj.*	合理的;理性的
anticipate	*vt.*	预期,期望;抢先
aerodynamics	*n.*	气体力学;航空动力学
resistance	*n.*	阻力;抵抗;反抗;抵抗力
aviation	*n.*	航空;飞行术;飞机制造业
steerable	*adj.*	易驾驶的;可操纵的

dirigible	*adj.* 可驾驶的
	n. 飞船
sporadical	*adj.* 零星的
sustain	*vt.* 维持;支撑,承担;忍受;证实
rigorous	*adj.* 严格的,严厉的;严密的;严酷的
aeronautics	*n.* 航空学;飞行术
clarify	*vt.* 澄清;阐明
	vi. 得到澄清;变得明晰;得到净化
drag	*vt.* 拖累;拖拉;缓慢而吃力地行进
	vi. 拖曳;缓慢而吃力地行进;阻力;阻力累
streamline	*vt.* 流线型化
curving	*adj.* 弯曲的
	n. 弯曲;曲线;变形
	v. 使弯曲;弯成弧形
comprise	*vt.* 包含;由……组成
fuselage	*n.* 机身(飞机)
glide	*vi.* 滑翔;滑行
	vt. 使滑行;使滑动
	n. 滑翔;滑行;滑移
methodological	*adj.* 方法的,方法论的
monoplane	*n.* 单翼机
biplane	*n.* 复翼飞机,双翼飞机
fabric	*n.* 组织;构造;建筑物
sleek	*adj.* 圆滑的;井然有序的
	vt. 使……光滑;掩盖
	vi. 打扮整洁;滑动
aluminum	*n.* 铝
autogyro	*n.* 旋翼机
rotor	*n.* 转子;水平旋翼;旋转体
powerplant	*n.* 动力装置

7

spun	*adj.* 纺成的
	v. 纺;旋转(spin 的过去分词)
propel	*vt.* 推进;驱使;激励;驱策
doctrine	*n.* 主义;学说;信条
coordinated	*adj.* 协调的
	vt. 调整;使调和;调节;整理(coordinate 的过去分词)
jet-powered	*adj.* 喷气动力的
cargo	*n.* 货物;船货;航空运输
glut	*vt.* 使……充满;使……吃饱;过多供应
	n. 供过于求;大量
usher	*n.* 引座员,带位员;接待员;门房
	vt. 引导,招待;迎接;开辟
	vi. 作招待员;当引座员
dubbed	*vt.* 刺;授予……称号;结账(dub 的过去分词)
	adj. 被称为的
moderate	*adj.* 稳健的,温和的;适度的,中等的;有节制的
	vi. 变缓和,变弱
	vt. 节制;减轻
countermeasure	*n.* 对策;反措施;反抗手段
supersonic	*adj.* 超音速的;超声波的
	n. 超音速;超声波
interceptor	*n.* 拦截机;妨碍者
autonomous	*adj.* 自治的;自主的;自发的
jump off	跳下来;开始,开始进攻
end in	以……为结果;以……告终
be attributed to	归因于……
low-powered	*adj.* 低功率的

III   Exercises

1   Translate the following phrases into English

(1) 固定翼飞机

(2) 地空导弹

(3) 无人机

(4) 点对点飞行

(5) 喷气时代

(6) 推重比

(7) 军事用途

(8) 比空气重的飞行

(9) 远距离轰炸机

(10) 运动第三定律

2   Translate the following sentences into Chinese

(1) Ballooning became a major "rage" in Europe in the late 18th century, providing the first detailed understanding of the relationship between the altitude and atmosphere.

(2) Conducting scientific aerodynamic experiment and demons and streamlining, movement of the center of pressure, and the increase in lift from curving the wing surface.

(3) Airplanes evolved from low-powered biplanes made from wood and fabric to sleek, high-powered monoplanes made of aluminum, primarily based on the founding work of Hugo Junkers during the World War I period and its adoption by American designer William Bushnell Stout and Soviet designer Andrei Tupolev.

(4) Even a moderate fleet of long-range bomber could deliver a deadly blow to the enemy, so great efforts were made to develop the countermeasure.

(5) Defining the modern aero plane configurati and comprising a fixed wing, fuselage and tail assembly.

IV   Supplement Reading

Area 51 Spy Plane and other Aviation tales

By Thom Patterson

It looks like an upside-down bathtub with wings, pretty odd for a spy jet that was among the nation's most highly classified pieces of military hardware.

As I stand in front of the plane code-named Tacit Blue at the National Museum of the U. S. Air Force, near Dayton, Ohio, I'm reminded that it still holds a bit of mystery.

Engineers made fun of Tacit Blue's design by nicknaming it the Whale, but the program—declassified in 1996—was deadly serious. It was all about stealth. Pentagon Cold War strategists desperately wanted to build planes that could evade Soviet radar.

And so the Air Force launched a "black program" to develop Tacit Blue and tested it at a secret government airbase in Nevada called "Area 51," according to CIA documents released in 2013.

The program, which lasted from 1978 to 1985, aimed to develop a single-seater jet for battlefield surveillance.

Before last year's document release, the government had never acknowledged the existence of Area 51. For decades, a fenced-off area surrounding Nevada's Groom Lake was rumored to be a testing ground for some of the nation's most secret technology.

Two retired Air Force test pilots who flew Tacit Blue in the early 1980s, Ken Dyson and Russ Easter, spoke about why this plane was important and what sets it apart.

Although the plane flew 135 times and was never put into production, without Tacit Blue, there would have been no B – 2 Spirit bomber. The plane proved that aircraft with curved surfaces could evade radar.

"The airplane flew pretty solid, I'd say," Dyson remembered.

Could sightings of Tacit Blue have contributed to UFO reports?

"I'm not aware of any circumstance like that," Easter said. Dyson also says no.

But Cynda Thomas, widow of the first Tacit Blue test pilot Richard G. "Dick" Thomas said she was with her husband in Los Angeles when an airline pilot accosted her husband during a test pilots' banquet at the Beverly Hilton.

As she remembers it, "The pilot came over, and he said, 'Mr. Thomas, I'm so-and-so, and I fly for Continental, and I'm sure I saw you flying the Tacit Blue—and you know, I reported you as a UFO.'"

"Airline pilots have, over the past, reported some stuff that could have been black aircraft in flight tests," Dyson said.

"One and a half of Tacit Blue planes were built", Dyson said, so that "if we lost one, we could have a second one up and flying in short order." What happened to the other half of Tacit Blue? "I think it was done away with—with total respect to secrecy."

Mechanical remnants from a related black program called "Have Blue" are buried at Groom Lake, according to a 2011 Air Force report. Groom Lake is inside Area 51, according to those released CIA documents.

"I don't know anything at all about that Have Blue stuff and wouldn't answer it if I did," said Dyson, who also tested Have Blue airplanes.

Dyson is aware of the CIA documents but said he didn't want to talk about Groom Lake or Area 51 or even to mention those places by name. "That's just because of the secrecy that was drilled into me," he said.

Maintaining Tacit Blue's secrets and preventing leaks, Dyson said, was proof of the success of a tightly knit and dedicated team. Pursuing a career centered around a secret job takes discipline.

"My wife had no clue what I was doing for a long time," Dyson said. "I just didn't talk about it to her or to anyone else who wasn't cleared on the program. It just wasn't done."

Richard Thomas also kept details about his work from his family, Cynda Thomas said, although in 1978 he did reveal to her that he was "going into the black world."

Secrecy makes professional relationships complicated at times. Associates outside Dyson's and Easter's secret circle wanted to hear "war stories" about what it was like to work in the world of black programs. Easter said, "Sometimes I wish we could tell more stories more freely so that some of the lessons learned could be passed on freer."

Three unique things about Tacit Blue

● Pantyhose made it safer: According to Cynda Thomas' book "Hell of a Ride", an air compressor was blowing tiny flammable aluminum shavings through inlet pipes and into the cockpit, creating a fire hazard. The engineers' unorthodox solution: "cover the inlet pipes with filters made of pantyhose," Thomas wrote. "That could be true," Easter acknowledges.

● They created an artificial wind tunnel with a huge transport aircraft. Developers used a huge C – 130 Hercules plane to create artificial winds that hit the side of the plane so they could test Tacit Blue's performance. Dyson described it as a sort of "wind tunnel" that was set up "in the black of the night so no one could see us flying overhead with a satellite."

● It was very hard to pilot: "Tacit Blue at the time was arguably the most unstable aircraft man had ever flown," ex-Northrop engineer John Cashen told Air Force Magazine.

Richard Thomas died in 2006, at age 76. Cynda Thomas said her husband had been battling Parkinson's disease.

At Tacit Blue's 1996 unveiling ceremony at the museum, Thomas was able to sit inside the airplane's cockpit one more time.

"That's when they let me point my camera up the steps of the airplane to take his picture," recalls Cynda Thomas. "I'm so thankful he got to do that before he passed. That plane was my husband's legacy."

The plane that wouldn't quit: "Spare 617"

Outside the museum's doors, at the Air Park, sit several giant planes—each with their own stories. One of them, the Air Force Museum says, saw "one of the greatest feats of airmanship of the Southeast Asia War."

In April 1972, a huge C130 – E Hercules transport plane code-named Spare 617 was ordered to fly over a raging battle in South Vietnam and parachute-drop giant pallets of ammunition[Figure 4]. If all went well, the ammo would resupply South Vietnamese soldiers fighting on the ground. However, all did not go well.

Figure 4   One of the Vietnam War's "greatest feats of airmanship" took place in 1972 aboard this C – 130E Hercules

Preparing to make his drop, pilot William Caldwell flew the plane low over the town of An Loc.

But the enemy had put a machine gun nest high above the town in a church steeple, Caldwell recalled, "So, we were just a sitting duck for him."

Caldwell remembered machine gun fire ripping through the cockpit, smashing a circuit breaker panel and the plane's windows. Flight engineer Jon Sanders died instantly. The attack wounded copilot John Hering and navigator Richard Lenz and damaged two of the plane's four turboprop engines.

Gunfire ruptured a duct designed to bleed hot air from the plane's powerful engines. The scalding air severely burned cargo loadmaster Charlie Shaub, as Caldwell put it, like a "600-degree hurricane."

Then it got worse. The attack set fire to some of the plane's explosive cargo. But despite its burn, Shaub was somehow able to eject the burning pallets of ammunition. It was just in time. Seconds later, the ammo exploded as it fell to the ground. Then Shaub astonishingly snuffed out the fire in the cargo hold. Caldwell closed the bleed air duct and shut down the damaged engines.

Next problem: how to save the wounded crew? Caldwell pointed the plane toward an air base with the best medical facilities. He would have to land a giant C – 130E with only two working engines—both on the same side of the aircraft.

"Things looked grim, but I got more confident with every mile we got closer to the air base," said Caldwell.

Again things went further south.

A hydraulics system that was needed to lower the landing gear became useless. Using nothing but sheer muscle, Lenz and cargo loadmaster Dave McAleece lowered the wheel by hand using crank handles, Caldwell said.

Caldwell landed the plane fast: pushing about 170 mph. With hydraulics busted, he had trouble steering the plane.

"I used the inboard right side engine to guide the plane down a high-speed runway turnoff," he said.

After rolling to a stop, "I got out of the airplane," Caldwell recalled. An airman on the ground asked, "Are you OK?" Caldwell replied, "You bastards didn't prepare us for this."

Blood Falls and other natural oddities.

Three notable details about the mission

• What recognition did the crew receive? Caldwell and Shaub received the Air Force Cross, the Air Force's second-highest award for valor.

• Did the plane have any weapons? Spare 617 flew with virtually no defensive weaponry. "We only had 38 revolvers," said Caldwell, 70, now a retired colonel who teaches aviation management at Southern Illinois University. It was the airplane itself, he said, that helped save them. "That airplane was just as responsible for getting us home as any of the crew."

• How hard was it to lower the landing gear? The attack knocked out the plane's hydraulic system, forcing the crew to lower the plane's gigantic landing gear manually. "I think it takes roughly 650 turns of that crank to get a landing gear down on both sides of the airplane. It's an endurance thing more than strength thing."

"Angel" of freedom: The Hanoi Taxi

A third piece of flying history at the museum had come to symbolize freedom for former U. S. POWs captured during the Vietnam War.

They called it the Hanoi Taxi, the first U. S. plane to ferry newly freed troops out of North Vietnam. Photos of the ex-captives taken aboard the C – 141 Starlifter clearly illustrate the joy and relief the men felt that day in February 1973[Figure 5].

Figure 5   American ex-POWs celebrate leaving North Vietnam in 1973 aboard a C – 141 Starlifter dubbed the Hanoi Taxi

Until the POWs left enemy airspace, they had refused to give their captors the satisfaction of showing happiness for being released. When they crossed over international waters, stoic silence immediately turned to joyful pandemonium.

A few hours later, with the Starlifter in the background, TV viewers watched stunning, emotional reunions in the Philippines between the ex-POWs and their loved ones.

Maj. Gen. Edward Mechenbier, a retired Air Force fighter pilot who spent nearly six years as a North Vietnamese prisoner, rode aboard the plane as it carried him back to the United States from the Philippines.

"This plane looked like an absolute angel coming to get us," he remembered.

Now the plane sits at the museum's Air Park, at the mercy of the wind and cold. Walking around the aircraft and running your hand across its metal exterior, you get a sense that you're touching a piece of history.

"It's a hallowed place," said museum curator Jeff Duford. "And you can definitely get a sense of that when you step aboard it. " But right now, that's not possible for visitors. The Taxi's interior is off-limits, and its flight deck windows are covered.

The plane—along with Spare 617—is slated to be housed in a new MYM35. 4 million building in 2016. Duford says the museum hasn't decided whether visitors will be able to board either aircraft when they move inside.

The Hanoi Taxi flew two freedom missions carrying ex-POWs out of Hanoi's Gia Lam Airport.

Overall, the aircraft's total passengers to freedom numbered 78 POWs and two civilian returnees. The plane also flew four freedom missions from the Philippines to the United States, carrying a total of 76 ex-POWs.

For four decades, the airplane served around the world, until it retired in 2006 to the Air Force museum.

Three unique things about the Hanoi Taxi

● The plane returned to Vietnam in 2004: Mechenbier, by then a major general, flew the Hanoi Taxi back to Vietnam on a mission to repatriate the newly recovered remain of two U. S. service members who were killed in action during the war.

● It served in the Iraq War: The plane helped transfer wounded troops from Baghdad to Ramstein Air Base in Germany and to Washington for treatment at Walter Reed Army Medical Center.

● It served as a disaster evacuation plane: The Hanoi Taxi airlifted survivors to safety after 2005's Hurricane Katrina.

Like many of the museum's other hundreds of aircrafts, these three planes—Tacit Blue, Spare 617 and the Hanoi Taxi—contributed to history.

"These really aren't airplanes anymore," Duford said. "They're artifacts. And so we want to make sure, as a responsible institution, that we protect those artifacts. That's more important than anything else."

Unit 2

Aircraft Structure

I   Reading Text

1   Introduction

When designing an aircraft, it's all about finding the optimal proportion of the weight of the vehicle and payload. It needs to be strong and stiff enough to withstand the exceptional circumstances in which it has to operate. Durability is an important factor. If a part fails, it doesn't necessarily result in failure of the whole aircraft.

The main sections of an aircraft, the fuselage, tail and wing, determine its external shape. The load bearing members of these main sections, those subjected to major forces, are called the airframe. The airframe is what remains if all equipment and systems are stripped away.

Old aircraft had skin made from impregnated linen that could hardly transmit any force at all. In most modern aircraft, the skin plays an important role in carrying load. Sheet metals can usually only support tension. But if the sheet is folded, it suddenly does have the ability to carry compressive load. Stiffeners are used for that. A section of skin, combined with stiffeners, called stringers, is termed a thin-walled structure.

A very good way of using sheet metal skin is in a thin-walled cylinder, called a monocoque structure. A cylinder with holes, for doors and such, is called a semi-monocoque structure. An extruded stiffener is manufactured by squeezing hot, viscous material through an opening of a certain shape. It can usually be recognized by the fact that the thickness is not consistent, especially in the corners. This is relatively expensive, compared to stiffener made from sheet metal. From sheet metal it is not possible to make complicated stiffeners. Thin sheet metal can be rolled or drawn.

Usually stiffeners are attached to the skin. In an integral structure, the skin and stiffeners have been manufactured from one solid block of material. It is also possible to make some kind of a sandwich structure, in which the skin has a high stiffness due to its special structure.

2　The fuselage

The fuselage should carry the payload and is the main body to which all parts are connected. It must be able to resist bending moment (caused by weight and lift from the tail), torsional load (caused by fin and rudder) and cabin pressurization. The structural strength and stiffness of the fuselage must be high enough to withstand these loads. At the same time, the structural weight must be kept to a minimum.

In transport aircraft, the majority of the fuselage is cylindrical or near-cylindrical, with tapered nose and tail sections. The semi-monocoque construction, which is virtually standard in all modern aircraft, consists of a stressed skin with added stringers to prevent buckling, attached to hoop-shaped frames.

The fuselage also has members perpendicular to the skin, that support it and help keep its shape. These supports are called frames if they are open or ring-shaped, or bulkheads if they are closed.

Disturbances in the perfect cylindrical shell, such as doors and windows, are called cutout. They are usually unsuitable to carry many of the load that are present on the surrounding structure. The direct load paths are interrupted and as a result, the structure around the cut-out must be reinforced to maintain the required strength.

3　Wing Content

Providing lift is the main function of the wings of an aircraft. The wings consist of two essential parts. The internal wing structure consists of spars, ribs and stringers, and the external wing is the skin.

Ribs give the shape to the wing section, support the skin (prevent buckling) and act to prevent the fuel surging around as the aircraft manoeuvres. They serve as attachment points for the control surfaces, flaps, undercarriage and engines. The ribs need to support the wing-panels, achieve the desired aerodynamic shape and keep it, provide points for conducting large forces, add strength, prevent buckling, and separate the individual fuel tanks within the wing.

The stringers on the skin panels run in the length of the wing, and so usually need to bridge the ribs. There are several methods for dealing with this problem. The stringers and ribs can both be uninterrupted. The stringers now run over the rib, leaving a gap between rib and skin. Rib and skin are indirectly connected, resulting in a bad shear load transfer between rib and skin. The stringers can be interrupted at the rib. Interrupting the stringer in this way certainly weakens the structure, and therefore extra strengthening material, called a doubler, is usually added. Naturally, the stringers can also interrupt the rib. The stringers now run through holes cut into the rib, which also causes inevitable weakening of the structure.

The ribs also need to be supported, which is done by the spars. These are simple beams that usually have a cross-section similar to an I-beam. The spars are the most heavily loaded parts of an aircraft. They carry much more force at its root, than at the tip. Since wings will bend upwards, spars usually carry shear force and bending moment.

4 The Tail

In most aircraft, the sole function of the tail unit is to provide the required stability and control. Stability is the tendency of the aircraft to return to its original attitude by itself.

Since an aircraft flies in three-dimensional space, stability and control are required in three direction. These axes are lateral (left and right), vertical (up and down) and longitudinal (fore and aft). For aircraft turn, three maneuver cases are used. For pitch, which is rotation about the lateral axis, the horizontal tail with elevators is used. For yaw, which is rotation about the vertical axis, the vertical fin with rudder is used. For roll, which is rotation about the longitudinal axis, the ailerons are used.

The fin provides stability in yaw. When the aircraft is required to yaw, the rudder is deflected. The tailplane provides stability in pitch. When the aircraft is required to climb or descend, the elevators are deflected.

If the position of the center of gravity varies, or the aircraft speed is changed, the elevator position necessary to maintain level flight will change. Therefore, a small extra control surface is added to each main surface to allow the pilot to trim the aircraft.

5 Undercarriage

The undercarriage of an aircraft support the aircraft on the ground, provide smooth taxiing and absorb shocks of taxiing and landing. It serves no function during flight, so it must be as small and light as possible, and preferably easily retractable.

Due to the weight of the fore and aft part of the aircraft, large bending moment occur on the center section. To carry these bending moments, a strong keel beam is present. This reduces the space in which the landing gear can be retracted.

When an aircraft lands, a large force is generated on the undercarriage as it meets the ground. To prevent damage to the structure, this shock must be absorbed and dissipated as heat by the undercarriage. If the energy is not dissipated, the spring system might just make the aircraft bounce up again.

After touchdown, the aircraft needs to brake. Disc brakes, which are primarily used, consist of a disc or a series of discs, gripped between pads of friction material. The braking of an aircraft can be supplemented by other forms of braking, such as air brake, causing a large increase in drag, or reverse thrust, thrusting air forward.

II   Words and Expressions

optimal	*adj.* 最佳的;最理想的
payload	*n.* 有效载荷;收费载重;工资负担
stiff	*adj.* 呆板的;坚硬的;严厉的;拘谨的;稠的
	adv. 极其;僵硬地;彻底地
	n. 死尸;令人讨厌者
	vt. 诈骗;失信
withstand	*vt.* 抵挡;禁得起;反抗

	vi. 反抗
exceptional	adj. 异常的,例外的
durability	n. 耐久性;坚固;耐用年限
fuselage	n. 机身
airframe	n. 机身
impregnated	adj. 浸渍的;浸染的
	vt. 浸渍(impregnate 的过去分词)
tension	n. 张力,拉力;紧张
	vt. 使紧张;使拉紧
compressive	adj. 压缩的;有压缩力的
stiffener	n. 加固物;加劲杆;刚性元件;兴奋剂
stringer	n. 纵梁
cylinder	n. 圆筒;汽缸;柱面;圆柱状物
monocoque	n. 单体横造;硬壳式构造
extruded	adj. 压出的;受挤压的
	vt. 使……喷出;使伸出;驱逐(extrude 的过去分词)
squeeze	vt. 挤;紧握;勒索
	vi. 压榨
	n. 压榨;紧握;拥挤
consistent	adj. 始终如一的,一致的;坚持的
integral	n. 积分;部分;完整
	adj. 积分的;完整的,整体的
moment	n. 力矩
torsional	adj. 扭转的;扭力的
rudder	n. 船舵;飞机方向舵
cabin	n. 小屋
viscous	adj. 黏性的;黏的
tapered	adj. 锥形的
	vt. 逐渐变少(taper 的过去式和过去分词)
buckle	n. 屈曲;膨胀;褶皱;下垂;粗糙度
	vt. 把……扣紧;连接
perpendicular	adj. 垂直的

bulkhead	*n.*	隔板;防水壁
disturbance	*n.*	干扰;失调;困惑
cutout	*n.*	排气阀;保险开关
internal	*n.*	内脏;本质
	adj.	内部的;里面的
spar	*n.*	帆桅杆;帆横杆(spar 的复数形式)
external	*n.*	外部;外观
	adj.	外部的;表面的
rib	*n.*	翼肋;肋状物
	vt.	戏弄;装肋于
surge	*n.*	汹涌;大浪,波涛
	v.	汹涌;起大浪,蜂拥而来
manoeuvre	*n.*	策略(同 maneuvre)
	vt.	诱使;操纵;耍花招
	vi.	调动;演习;用策略
flap	*n.*	襟翼
beam	*n.*	梁
three-dimensional	*adj.*	三维的;立体的;真实的
shear	*vt.*	剪;修剪;剥夺
	vi.	剪;剪切;修剪
	n.	切变;修剪;大剪刀
axe	*n.*	斧
	vt.	削减;用斧砍
pitch	*vi.*	倾斜;投掷;坠落
	vt.	投;掷;定位于;用沥青涂;扎营;向前倾跌
	n.	沥青;程度;树脂;倾斜;投掷
rotation	*n.*	旋转;循环,轮流
lateral	*n.*	侧部
	adj.	侧面的,横向的
	vt.	横向传球
yaw	*n.*	偏航
	vt.	使……偏航

	vi. 偏航
aileron	*n.* 副翼
deflect	*vi.* 打歪(deflect 的过去分词)
	adj. 偏离的,偏向的
elevator	*n.* 升降舵
retractable	*adj.* 可取消的,可缩回的
aft	*adv.* 在船尾;近船尾
keel	*vt.* 给……装龙骨;把……翻转
	n. 龙骨;平底船;龙骨脊
	vi. 翻倒;倾覆
undercarriage	*n.* 飞机起落架,着陆装置;底盘
dissipate	*vt.* 浪费;使……消散
	vi. 驱散;放荡
drag	*n.* 阻力
thrust	*n.* 推力
thin-walled	*adj.* 薄壁的
ring-shaped	*adj.* 环形的;环状的;圆形的
run through	跑着穿过;浏览;刺

Ⅲ Exercises

1 Translate the following phrases into English

(1)断流器
(2)油箱
(3)横截面
(4)剪切力
(5)弯矩

（6）龙骨梁

（7）单壳机身

（8）三维空间

（9）重心

（10）弹起

2   Translate the following sentences into Chinese

（1）A very good way of using sheet metal skin is in a thin-walled cylinder, called a monocoque structure.

（2）An extruded stiffener is manufactured by squeezing hot, viscous material through an opening of a certain shape.

（3）The stringers now run through holes cut into the rib, which also causes inevitable weakening of the structure.

（4）Due to the weight of the fore and aft part of the aircraft, large bending moment occurs on the center section.

（5）In most aircraft, the sole function of the tail unit is to provide the required stability and control.

IV   Supplement Reading

Aircraft Components & Structure

Introduction

- The airframe is the basic structure of an aircraft and is designed to withstand all aerodynamic forces, as well as the stresses imposed by the weight of the fuel, crew, and payload

- Although similar in concept, aircraft can be broken down into fixed and rotary wing structures
- The airplane is controllable around its lateral, longitudinal, and vertical axes by deflection of flight control surface
- These control devices are hinged or movable surfaces with which the pilot adjusts the airplane's attitude during takeoff, flight maneuvering, and landing
- They are operated by the pilot through connecting linkage by means of rudder pedal and a control stick or wheel
- Principle Structure
  - Fuselage

    main structural unit
  - Wings

    airfoils to produce lift
  - Flight Control Surface
    - Primary

      aileron, elevator, rudder
    - Secondary

      movable trim tabs located on the primary flight control surfaces
    - Auxiliary

      wing flaps, spoilers, speed brakes and slats

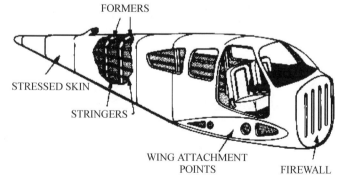

Figure 1　Aircraft Fuselage

Fuselage

- The fuselage is the principal structural unit of an aircraft
- The fuselage is designed to accommodate the crew, passengers, cargo, instruments, and other essential equipment
- Types of Fuselage Construction [Figure 1]
  - The construction of aircraft fuselages evolved from the early wood truss structural arrangements to monocoque shell structures and the current semimonocoque shell structures
    - Truss Structure

      In this construction method, strength and rigidity are obtained by joining tubing (steel or aluminum) to produce a series of triangular shapes, called trusses
    - Lengths of tubing, called longerons, are welded in place to form a wellbraced framework
    - Vertical and horizontal struts are welded to the longerons and give the structure a square or rectangular shape when viewed from the end
    - Additional struts are needed to resist stress that can come from any direction
    - Stringers and bulkheads, or formers, are added to shape the fuselage and support the covering
    - As designs progressed, these structures were enclosed firstly with cloth and eventually with metals
    - These upgrades streamlined shape and increased performance
    - In some cases, the outside skin can support all or a major portion of the flight loads
  - Most modern aircraft use a form of this stressed skin structure known as monocoque or semimonocoque construction [Figure 2]
    - Monocoque
    - (1) Monocoque (French for "single shell") construction uses stressed skin to support almost all loads much like an aluminum beverage can
    - (2) In monocoque construction, rigs, formers, and bulkheads of varying sizes give shape and strength to the stressed skin fuselage

(3) Despite very strong, monocoque construction is not highly tolerant to deformation of the surface

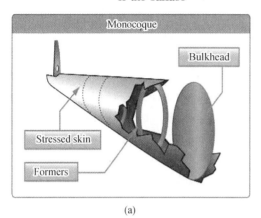
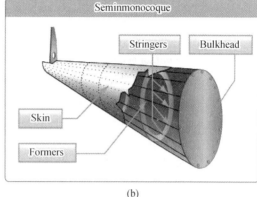

(a)   (b)

Figure 2   Monocoque fuselage and Semi-monocoque fuselage

(4) For example, an aluminum beverage can supports considerable forces at the ends of the can, but if the side of the can is deformed slightly while supporting a load, it collapses easily

(5) Because most twisting and bending stresses are carried by the external skin rather than by an open framework, the need for internal bracing was eliminated or reduced, saving weight and maximizing space

(6) One of the notable and innovative methods for using monocoque construction was employed by Jack Northrop

(7) In 1918, he devised a new way to construct a monocoque fuselage used for the Lockheed S – 1 Racer

(8) The technique utilized two molded plywood half-shells that were glued together around wooden hoops or stringers

(9) To construct the half shells, rather than gluing many strips of plywood over a form, three large sets of spruce strips were soaked with glue and laid in a semi-circular concrete mold that looked like a bathtub

(10) Then a rubber balloon was inflated in the cavity to press the plywood against the mold under a tightly clamped lid

(11) Twenty-four hours later, the smooth half-shell was ready to be joined to another to create the fuselage

(12) The two halves were each less than a quarter inch thick

(13)　Although employed in the early aviation period, monocoque construction would not reemerge for several decades due to the complexities involved

(14)　Every day examples of monocoque construction can be found in automobile manufacturing where the unibody is considered standard in manufacturing

- Semimonococque

Semimonococque construction, partial or one-half, uses a substructure to which the airplane's skin is attached. The substructure, which consists of bulkheads and/or formers of various sizes and stringers, reinforces the stressed skin by taking some of the bending stress from the fuselage. The main section of the fuselage also includes wing attachment points and a firewall. On the single-engine airplane, the engine is usually attached to the front of the fuselage. There is a fireproof partition between the rear of the engine and the flight deck or cabin to protect the pilot and passengers from accidental engine fire. This partition is called a firewall and is usually made of heat-resistant material such as stainless steel. However, a new emerging process of construction is the integration of composites or aircraft entirely made of composites

Wings

- Wings are airfoils attached to each side of the fuselage and are the main lifting surfaces that support the airplane in flight
- Wings may be attached at the top ("high-wing"), middle ("mid-wing"), or lower ("low-wing") portion of the fuselage
- The number of wings can also vary
  ○ Airplanes with a single set of wings are referred to as monoplanes, while those with two sets are called biplanes [Figure 3]

Figure 3　Monoplane and Biplane

- Many high-wing airplanes have external braces, or wing struts that transmit the flight and landing loads through the struts to the main fuselage structure [Figure 4]
- Since the wing struts are usually attached approximately halfway out on the wing, this type of wing structure is called semi-cantilever

SEMI-CANTILEVER WING   FULL-CANTILEVER WING

Figure 4   Semi-cantilever wing and cantilever wing

- A few high-wing and most low-wing airplanes have a full cantilever wing designed to carry the loads without external struts
- The principal structural parts of the wing are spars, ribs, and stringers [Figure 5]

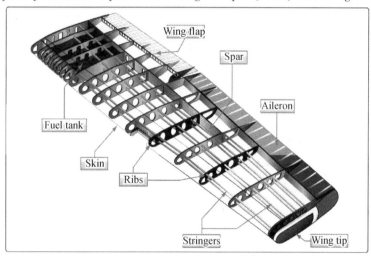

Wing flap
Spar
Aileron
Fuel tank
Skin
Ribs
Stringers
Wing tip

Figure 5   Wing Construction

- These are reinforced by trusses, I-beams, tubing, or other devices, including the skin
- The wing ribs determine the shape and thickness of the wing (airfoil)
- In most modern airplanes, the fuel tanks are either an integral part of the wing's structure or consist of flexible containers mounted inside of the wing
- Attached to the rear, or trailing edges, of the wings are two types of control surfaces referred to as ailerons and flaps

29

- Alternate Types of Wings
  - These design variations are related to Aerodynamics, which provides information on the effect controls have on lifting surfaces from traditional wings to wings that use both flexing (due to billowing) and shifting (through the change of the aircraft's CG). For example, the wing of the weight-shift control aircraft is highly swept in an effort to reduce drag and allow for the shifting of weight to provide controlled flight. Handbooks specific to most categories of aircraft are available for the interested pilot and can be found on the Federal Aviation Administration (FAA) website at www. faa. gov.
- Ailerons
  - Ailerons (French for "little wing") are control surfaces on each wing which control the aircraft about its longitudinal axis allowing the aircraft to "roll" or "bank"
    - This action results in the airplane turning in the direction of the roll/bank
    - With aileron deflection, there is asymmetrical lift (rolling moment) about the longitudinal axis and drag (adverse yaw)
  - They are located on the trailing (rear) edge of each wing near the outer tips
    - They extend from about the midpoint of each wing outward toward the tip, and move in opposite directions to create aerodynamic forces that cause the airplane to roll
  - The yoke manipulates the airfoil through a system of cables and pulleys and act in an opposing manor
    - Yoke "turns" left: left aileron rises, decreasing camber and angle of attack on the right wing which creates downward lift
    - At the same time, the right aileron lowers, increasing camber and angle of attack which increases upward lift and causes the aircraft to turn left
    - Yoke "turns" right: right aileron rises decreasing camber and angle of attack on the right wing which creates downward lift
    - At the same time, the left aileron lowers, increasing camber and angle of attack on the left wing which creates upward lift and causes the aircraft to turn

right

○ Although uncommon, some ailerons are configured with trim tabs which relieve pressure on the yoke on the aileron for rolling

- Wing Planform

  ○ The shape and design of a wing is dependent upon the type of operation for which an aircraft is intended and is tailored to specific types of flying [Figure 6].

    ■ Rectangular

    (1) Rectangular wings are best for training aircraft, as well as low speed aircraft

    (2) Designed with twist to stall at the wing root first, to provide aileron control in stalls

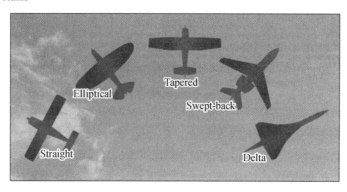

Figure 6   Airplane Flying Handbook, Airfoil types

    ■ Elliptical

      Elliptical wings are most efficient, but difficult to produce (spitfire)

    ■ Tapered

    ■ More efficient than a rectangle wing but easier to produce than an elliptical design

    ■ Swept

    (1) Usually associated with swept-back, but can also be swept-foreward

    (2) Swept-back wings are best for high speed aircraft for delaying Mach tendencies

    (3) Stall at the tips first, providing poor stall characteristics

    ■ Delta

    (1) Advantages of a swept wing, with good structural efficiency and low frontal area

31

(2) Disadvantages are the low wing loading and high wetted area needed to obtain aerodynamic stability

Empennage

- Commonly known as the "tail section," the empennage includes the entire tail group which consists of fixed surfaces such as the vertical fin or stabilizer and the horizontal stabilizer; the movable surfaces including the rudder and rudder trim tabs, as well as the elevator and elevator trim tabs[Figure 7]

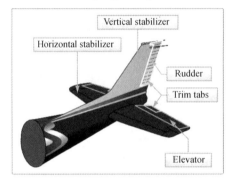

Figure 7　Empennage

- These movable surfaces are used by the pilot to control the horizontal rotation (yaw) and the vertical rotation (pitch) of the airplane
- In some airplanes the entire horizontal surface of the empennage can be adjusted from the cockpit as a complete unit for the purpose of controlling the pitch attitude or trim of the airplane. Such designs are usually referred to stabilators, flying tails, or slab tails
- Then the empennage provides the airplane with directional and longitudinal balance (stability) as well as a means for the pilot to control and maneuver the airplane
- Rudder
  ○ Rudders are used to control the direction (left or right) of "yaw" about an airplane's vertical axis
  ○ Like the other primary control surfaces, the rudder is a movable surface hinged to a fixed surface that, in this case, is the vertical stabilizer or fin
  ○ Its action is very much like that of the elevators, except that it swings in a different

plane-from side to side rather than up and down

- It is not used to make the airplane turn, as is often erroneously believed
- In practice, both aileron and rudder control input are used together to turn an aircraft, the ailerons imparting roll
- This relationship is critical in maintaining coordination or creating a slip
- Improperly ruddered turn at low speed can precipitate a spin

  ○ Rudders are controlled by the pilot with his/her feet through a system of cables and pulleys
    - "Step" on the right rudder pedal: rudder moves right creating a yaw to the right
    - "Step" on the left rudder pedal: rudder moves left creating a yaw to the left

- Elevator
  ○ The elevator, which is attached to the back of the horizontal stabilizer, is used to move the nose of the airplane up and down during flight
- Stabilator
  ○ A second type of empennage design does not require an elevator
  ○ On the contrary, it incorporates a one-piece horizontal stabilizer that pivots from a central hinge point
  ○ This type of design is called a stabilator and is moved using the control wheel, just as the elevator is moved
  ○ For example, when a pilot pulls back on the control wheel, the stabilator pivots so the trailing edge moves up
  ○ This increases the aerodynamic tail load and causes the nose of the airplane to move up. Stabilators have an anti-servo tab extending across their trailing edge [Figure 8]
  ○ The anti-servo tab moves in the same direction as the trailing edge of the stabilator and helps make the stabilator less sensitive
  ○ The anti-servo tab also functions as a trim tab to relieve control pressures and helps maintain the stabilator in the desired position

Figure 8   Stabilators

Flight Control Surfaces

- Flight control surfaces consist of primary, secondary, and auxiliary controls [Figure 9]

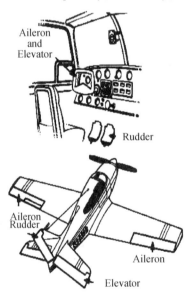

Figure 9   Flight Control Surfaces

- Auxiliary Flight Control Surfaces
    - Tabs are small, adjustable aerodynamic devices on the trailing edge of the control surface
    - These movable surfaces reduce pressures on the controls

o   Trim controls a neutral point, like balancing the aircraft on a pin with unsymmetrical weight

o   This is done either by trim tabs (small movable surfaces on the control surface) or by moving the neutral position of the entire control surface all together

o   These tabs may be installed on the ailerons, the rudder, and/or the elevator

o   Trim Tabs

  ▪   The force of the airflow striking the tab causes the main control surface to be deflected to a position that corrects the unbalanced condition of the aircraft

  ▪   When disturbed, an aircraft trimmed properly will try to return to its previous state due to aircraft stability

  ▪   Trimming is a constant task required after any power setting, airspeed, altitude, or configuration changes

  ▪   Proper trimming decreases pilot workload allowing attention to be diverted elsewhere, especially important for instrument flying

  ▪   Trim tabs are controlled through a system of cables and pulleys

  ▪   Trim tab adjusted up: trim tab lowers creating positive lift, lowering the nose This movement is very slight

  ▪   Trim tab adjusted down: trim tab raises creating positive lift, raising the nose This movement is very slight

  ▪   To learn more about how to use the trim tab in flight see the trimming of the aircraft

o   Servo Tabs

  ▪   Servo tabs are similar to trim tabs in that they are small secondary controls which help reduce pilot workload by reducing forces

  ▪   The defining difference, however, is that these tabs operate automatically, independent of the pilot

  ▪   Types of Servo Tab Designs:

  (1)   Anti-servo

      Also called an anti-balance tab, are tabs that move in the same direction as the control surface

  (2)   Servo

      Tabs that move in the opposite direction as the control surface

Controls

- Slats
  - Slats are part of the Flight Control System
  - Attached to the leading edge of the wings and are designed to be controlled by the pilot or automatically by the flight computer
  - Some aircraft employ aerodynamic slats that when lowered create added lift
  - Slats increase the camber of the wings/airfoil
  - By extending the slats additional lift is created when the aircraft is at slower airspeeds, normally on takeoff and landing
- Flaps
  - Flaps are part of the Flight Control System
  - Attached to the trailing edge of the wings and are controlled by the pilot from the cockpit
  - By extending the flaps additional lift is created when the aircraft is at slower airspeeds, normally on takeoff and landing
  - Slats and flaps are used in conjunction with each other to increase both lift and stall margin by increasing the overall wings camber thus, allowing the aircraft to maintain control flight at slower airspeeds
  - Flaps extend outward from the fuselage to near the midpoint of each wing
  - The flaps are normally flush with the wing's surface during cruising flight
  - When extended, the flaps move simultaneously downward to increase the lifting force of the wing for takeoffs and landings [Figure 8]
- Elevator
  control surfaces which control the aircraft about its lateral axis allowing the aircraft to pitch
  - The elevators are attached to the horizontal portion of the empennage—the horizontal stabilizer
    - The exception to this is found in those installations where the entire horizontal surface is a one piece structure which can be deflected up or down to provide longitudinal control and trimming

- A change in position of the elevators modifies the camber of the airfoil, which increases or decreases lift
- When forward pressure is applied on the controls, the elevators move downward
- This increases the lift produced by the horizontal tail surfaces
- The increased lift forces the tail upward, causing the nose to drop
- Conversely, when back pressure is applied on the wheel, the elevators move upward, decreasing the lift produced by the horizontal tail surfaces, or maybe even producing a downward force
- The tail is forced downward and the nose up
- The elevators control the angle of attack of the wings
- When back-pressure is applied on the controls, the tail lowers and the nose rises, increasing the angle of attack
- Conversely, when forward pressure is applied, the tail raises and the nose lowers, decreasing the angle of attack
- Stabilizer: a control surface other than the wings which provide stabilizing qualities

- Speed Brakes
  - Designed to slow the aircraft when in a dive or descent, location and style vary with aircraft, and are controlled by a switch in the cockpit
- Trim Tabs
  - Movable tabs located on the primary control surfaces i. e., ailerons, elevators and rudder reducing the pilot's workload enabling the aircraft to hold a particular attitude without the need of constant pressure/inputs into the system
- Landing Gear
  - The landing gear is the principal support of the airplane when parked, taxiing, taking off, or landing
  - A steerable nosewheel or tailwheel permits the airplane to be controlled throughout all operations while on the ground
  - Most aircraft are steered by moving the rudder pedals, whether nosewheel or tailwheel. Additionally, some aircraft are steered by differential braking
- Power Plant
  - The powerplant usually includes both the engine and the propeller
  - Engine

- The primary function of the engine is to provide the power to turn the propeller
- It also generates electrical power, provides a vacuum source for some flight instruments, and in most single-engine airplanes, provides a source of heat for the pilot and passengers [Figure 10]
- On single engine airplanes the engine is usually attached to the front of the fuselage
- There is a fireproof partition between the rear of the engine and the cockpit or cabin to protect the pilot and passengers from accidental engine fire. This partition is called a firewall and is usually made of a high heat resistant, stainless steel

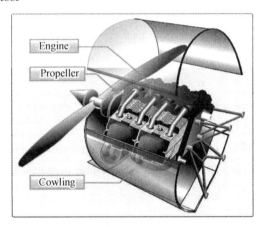

Figure 10   Aircralt Engine

○ Cowling
  - The engine is covered by a cowling, or a nacelle, which are both types of covered housing
  - The purpose of the cowling or nacelle is to streamline the flow of air around the engine and to help cool the engine by ducting air around the cylinders
○ Propeller
  - The propeller, mounted on the front of the engine, translates the rotating force of the engine into thrust, a forward acting force that helps move the airplane through the air

- A propeller is a rotating airfoil that produces thrust through aerodynamic action
- A high-pressure area is formed at the back of the propeller's airfoil, and low pressure is produced at the face of the propeller, similar to the way lift is generated by an airfoil used as a lifting surface or wing
- This pressure differential develops thrust from the propeller, which in turn pulls the airplane forward
- Engines may be turned around to be pushers with the propeller at the rear
- There are two significant factors involved in the design of a propeller that impact its effectiveness
- The angle of a propeller blade, as measured against the hub of the propeller, keeps the angle of attack (AOA) (see definition in Glossary) relatively constant along the span of the propeller blade, reducing or eliminating the possibility of a stall
- The amount of lift being produced by the propeller is directly related to the AOA, which is the angle at which the relative wind meets the blade
- The AOA continuously changes during the flight depending upon the direction of the aircraft
- The pitch is defined as the distance a propeller would travel in one revolution if it were turning in a solid
- These two factors combine to allow a measurement of the propeller's efficiency
- Propellers are usually matched to a specific aircraft/powerplant combination to achieve the best efficiency at a particular power setting, and they pull or push depending on how the engine is mounted

Rotary-Wing Components

- The major difference between helicopters and fixed-wing is the source of lift
- Fixed-winged aircraft derive lift from fixed airfoils while helicopters use rotating airfoils known as rotor blades
- Lift and control are relatively independent of forward speed
- Controls
  - Cyclic Stick

- Controls movement about the lateral and longitudinal axis of the helicopter
- It is located centered in front of the pilot's seat and changes the tip path plane of the main rotor for directional flight
- By changing the tip path plane, the direction of thrust is changed, and the corresponding intended direction of movement or flight is achieved
  - Collective Stick
    - Always located to the left of the pilot's seat and varies the lift of the main rotor by decreasing or increasing the angle of attack on all rotor plates equally and in the same direction
    - Also used in combination with the cyclic to regulate speed and altitude
  - Rudder Pedals
    - Controls movement about the vertical axis (yaw) of the helicopter by changing the pitch (angle of attack) of the tail rotor plates
    - This causes more or less force to be developed which is counteracting the torque caused by the main rotors
    - Additionally, by the pilot deflecting the rudder pedals left or right the aircraft heading or direction is changed left or right
- Components
  - Rotor Blade
    - Spinning "wings" which allow for lift on helicopters or "rotor-craft"
  - Main Rotor Assembly
    - Consists of rotor blades, rotor hub assembly, pitch control rod/links, mast, swashplate and support assembly
    - Some may have scissor and sleeve assembly
    - All of the above items work to change linear (push/pull motion) into rotating control movement
  - Gearboxes/Transmission
    - Changes direction and provides power produced by the engines via drive shafts to the main and trail rotor assemblies
    - The main transmission also provides mounting pads for accessory mounting such as hydraulic flight control pumps, generators, and rotor brake
    - Most helicopters have a main, intermediate and a tail gearbox

Conclusion

- The principles of flight are those basic characteristics which act upon an aircraft
- A balanced aircraft is a happy aircraft (fuel burn, efficiency, etc.)
- As aircraft construction evolved from truss truss structures, which lacked a streamlined shape, to the more formed monocoque and semimonocoque designs of today

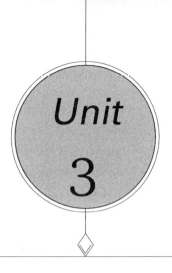

Unit 3

Air Traffic Control

I Reading Text

An accident that occurred in the skies over the Grand Canyon in 1956 resulted in the establishment of the Federal Aviation Administration (FAA) to regulate and oversee the operation of aircraft in the skies over the United States, which were becoming quite congested. The resulting structure of air traffic control procedures are also in place over much of the rest of the world.

Rudimentary air traffic control (ATC) existed well before the Grand Canyon disaster. As early as the 1920s, the earliest air traffic controllers manually guided aircraft in the vicinity of the airports, using lights and flags, while beacons and flashing lights were placed along cross-country routes to establish the earliest airways. However, this purely visual system was useless in bad weather, and by the 1930s, radio communication was coming into use for ATC. The first region to have something similar to today's ATC was New York City, with other major metropolitan areas following soon after.

In the 1940s, ATC centers could take advantage of the newly developed radar and improved radio communication brought about by the Second World War, but the system remained rudimentary. It was only after the creation of the FAA that full-scale regulation of American airspace took place, and this was fortuitous, for the advent of the jet engine suddenly resulted in a large number of very fast planes, reducing pilots' margin of error and practically demanding some set of rules to keep everyone well separated and operating safety in the air.

Many people think that ATC consists of a row of controllers sitting in front of their radar screens at the nation's airports, telling arriving and departing traffic what to do. This is a very incomplete part of the picture. The FAA realized that the airspace over the United States would at any time have many different kinds of planes, flying for many different purposes, in a variety of weather conditions, and the same kind of structure was needed to accommodate all of them.

To meet this challenge, the following elements were put into effect. First, ATC extends over virtually the entire United States. In general, from 365 m above the ground and higher, the entire country is blanketed by controlled airspace. In certain areas, mainly near airports, controlled airspace in which FAA regulations apply. Elsewhere, in uncontrolled airspace, pilots are bound by fewer regulations. In this way, the recreational pilot who simply wishes to go flying for a while without all the restrictions imposed by the FAA has only to stay in uncontrolled airspace, below 365 m, while the pilot who does want the protection afforded by ATC can easily enter the controlled airspace.

The FAA then recognized two types of operating environments. In good meteorological conditions, flying would be permitted under Visual Flight Rules (VFR), which suggests a strong reliance on visual cues to maintain an acceptable level of safety. On a clear day, a pilot in controlled airspace can choose a VFR or IFR flight plan, and the FAA regulations were devised in a way which accommodates both VFR and IFR operations in the same airspace. However, a pilot can only choose to fly IFR if they possess an instrument rating which is above and beyond the basic pilot's license that must be held.

We divided controller airspace into several different types, designated by letters of the alphabet. Uncontrolled airspace is designated Class F, while controlled airspace below 5,490 m above sea level and not in the vicinity of an airports is Class E. All airspace above 5,490 m is designated Class A. The reason for the division of Class E and Class A airspace stems from the types of planes operating in them. Generally, Class E airspace is where one finds general aviation aircraft(few of which can climb above 5,490 m), and commercial turboprop aircraft. Above 5,490 m is the realm of the heavy jet, since jet engines operate more efficiently at higher altitudes. The difference between Class E and Class A airspace is that in Class A, all operations are IFR, and pilots must be instrument-rated, that is, skilled and licensed in aircraft instrumentation. This is because ATC control of the entire space is essential. Three other types of airspace, Class D, C and B, govern the vicinity of airports. These correspond roughly to small municipal, medium-sized metropolitan and major metropolitan airports respectively, and encompass an increasingly rigorous set of regulations. For example, all a VFR pilot has to do to enter Class C airspace and establishes two-way radio contact with ATC. No explicit permission from ATC to enter is needed, although the pilot must continue to obey all regulations governing VFR flight. To enter Class B airspace, such as on approach to a major metropolitan airport, an explicit ATC clearance is required. The private pilot who cruises without permission into this airspace risks losing their license.

Every sector within an ARTCC usually has one to three controllers assigned to separate the aircraft within that sector. The first position that most controllers in an ARTCC are assigned to is the role of flight data controller. The flight data controller is responsible for assisting the other controllers, who actually separate the aircraft. The flight data controller effects coordination with other controllers and passes along pertinent flight information to controllers working in other sectors.

Every ARTCC sector equipped with radar is staffed by a controller whose responsibility is to separate participating aircraft using a radar-derived display. Radar controllers issue altitude, heading, or airspeed changes to keep the aircraft separated and be in compliance with the various letters of agreement and facility directives that may apply to that sector.

Every sector within the center is also staffed by a radar associate/nonradar controller whose duties are to assist the radar controller when separating aircraft that do not appear on the radar display. The nonradar controller's duties include updating the flight progress strips to accurately reflect every aircraft's position, altitude, and route of flight. The nonradar controller uses this information to separate aircraft that are either too low or too far away to be displayed on the radar. The nonradar controller must be prepared to assume aircraft separation responsibility if the radar display should malfunction. The nonradar controller's duties are similar to those performed by the B controller in the old air traffic control centers.

When it is operationally advantageous for an ARTCC to delegate separation responsibility to an air traffic control tower (ATCT), an appropriate letter of agreement is drafted by representatives of both the tower and the center. This letter of agreement delineates the control tower's area of responsibility and formally transfers the responsibility for aircraft separation to the tower. In most cases, the control tower is delegated the responsibility for separation of participating aircraft operating within about a 40-mile radius of the airport. This airspace usually extends from the Earth's surface up to an altitude of 6,000 to 10,000 feet MSL.

The ground controller works in the glass-enclosed portion of the tower known as the tower cab and is responsible for the separation of aircraft and vehicles operating on the ramp, taxiways, and any inactive runways. This responsibility includes aircraft taxiing out for takeoff, aircraft taxiing into the terminal building after landing, and any ground vehicles operating on airport movement areas. Airport movement areas do not include those areas solely reserved for vehicular traffic such as service roads or boarding areas.

The ground controller is assigned a unique radio frequency to communicate with pilots and vehicle operators. The most common ground control frequency is 121.90 MHz. In congested areas where two or more control towers are located near each other, ground controller transmissions from each airport might overlap, causing pilot misinterpretation. Thus in such cases each control tower is assigned a different frequency for its ground controllers. These additional frequencies are usually 121.80 MHz or 121.70 MHz.

The duties of the ground controller include:

(1) Providing instructions to taxiing aircraft and ground support vehicles.

(2) Controlling taxiway lighting systems.

(3) Issuing clearances to IFR and participating VFR aircraft.

(4) Coordinating with the local controller when taxiing aircraft need to operate on active runways.

(5) Issuing weather and NOTAM information to taxiing aircraft.

(6) Receiving and relaying IFR departure clearances.

(7) Relaying runway and taxiway condition information to airport management.

At less busy air traffic control towers, the ground controller may also be responsible for coordinating with other facilities and issuing ATC clearances to aircraft prior to departure. At busier control towers, these tasks are assigned to a clearance delivery controller, who is assigned a frequency separate from that used by the ground controller. At more busy locations, a flight data controller may also be on duty to assist the ground controller when coordinating with other controllers.

The local controller is primarily responsible for the separation of aircraft operating within the airport traffic area and those landing on any of the active runways. The local controller is assigned to a unique radio frequency that permits communication with these aircrafts. The primary responsibility of the local controller is arranging inbound aircraft into a smooth and orderly flow of traffic and sequencing departing aircraft into this flow. The local controller's responsibilities are complicated by the fact that most of the airports in this country do not have sufficient nonintersecting runways to handle the number of aircraft that want to land or take off. Thus the local controller may be forced to use two or three runways that intersect each other.

At very busy facilities, the local controller's workload may be too large for one person to handle. In such cases, the local control position is split into two, with each controller responsible for different runways and assigned separate radio frequencies. Duties performed by the local controller include:

(1) Determining the active runway.

(2) Issuing landing and takeoff clearances.

(3) Issuing landing information.

(4) Sequencing landing aircraft.

(5) Coordinating with other controllers.

(6) Issuing weather and NOTAM information to pilots.

(7) Operating the runway and approach light systems.

II Words and Expressions

congested	adj. 堵塞的,拥挤的
rudimentary	adj. 基本的;初步的
manually	adv. 手动地;用手
vicinity	adv. 附近
metropolitan	n. 大都会
fortuitous	adj. 偶然的
margin	n. 保证金
cue	n. 提示
instrument	n. 仪器
Recreational	n. 休闲
designated	adj. 指定的
alphabet	n. 字母

turboprop	*n.* 涡轮螺旋桨飞机
realm	*n.* 领域
govern	*vt.* 管理
municipal	*n.* 市政
encompass	*vt.* 包含
explicit	*adj.* 精确的
full-scale	*adj.* 全面的
nonintersect	*vi.* 不相交
delineate	*n.* 要表达的观点
meet this challenge	迎接这一挑战

Ⅲ Exercises

1 Translate the following phrases into English

(1) 无线电通信

(2) 生效

(3) 管制空域

(4) 目视飞行规则

(5) 在附近

(6) 符合

(7) 玻璃幕墙

(8) 塔台工作间

(9) 活跃跑道

(10) 进港飞行器

2 Translate the following sentences into Chinese

1. As early as the 1920s, the earliest air traffic controllers manually guided aircraft in the vicinity of the airports, using lights and flags, while beacons and flashing lights were placed along cross-country routes to establish the earliest airways.

2. It was only after the creation of the FAA that full-scale regulation of America's airspace took place, and this was fortuitous, for the advent of the jet engine suddenly resulted in a large number of very fast planes, reducing pilots' margin of error and practically demanding some set of rules to keep everyone well separated and operating safety in the air.

3. In this way, the recreational pilot who simply wishes to go flying for a while without all the restrictions imposed by the FAA has only to stay in uncontrolled airspace, below 365 m, while the pilot who does want the protection afforded by ATC can easily enter the controlled airspace.

4. When it is operationally advantageous for an ARTCC to delegate separation responsibility to an air traffic control tower (ATCT), an appropriate letter of agreement is drafted by representatives of both the tower and the center.

5. The ground controller works in the glass-enclosed portion of the tower known as the tower cab and is responsible for the separation of aircraft and vehicles operating on the ramp, taxiways, and any inactive runways.

IV Supplement Reading

How They Stop Planes Colliding On World's Busiest Runways

As they stare out of the four-story plate glass windows of Heathrow's control tower, Jason Cooper and his team should be gnawing their fingernails and swabbing the sweat from their brows.

On any given day, they're in control of one of the most insanely busy stretches of airport runway on the planet, guiding tens of thousands of people on and off the ground.

One false move could spell disaster: at best the costly loss of a precious runway slot; at worst a genuine aviation tragedy and massive disruption across European airspace.

This is why it's both surprising and reassuring to find, during a recent CNN visit, that the atmosphere inside the tower at Heathrow, the UK's largest airport, is one of almost Zen-like calm.

"There's pressure, but no stress," says Cooper, who manages Red Watch, one of three groups of air traffic controllers (ATCs) that staff Heathrow's tower day and night. "I think most of us enjoy the pressure."

Under his command today are an apparently genial mixture of men and women of varying ages, all dressed down.

They do their jobs quietly and efficiently, talking into radio headsets, occasionally chatting to one another.

Outside though, the activity is relentless.

High wire balancing act

Heathrow control tower stands 87 meters over the airport[Figure 1].

Figure 1   Heathrow Tower

With five terminals and only two runways—each handling 19 million more passengers annually than any other airport in the world—Heathrow is perpetually congested.

Airliners, including giant Airbus A380s, line up on taxiways for takeoff, while the sky is filled with an endless parade of jets coming in from or heading out to destinations around the globe.

It's the job of the tower controllers to keep snarl-ups to a minimum—a high wire balancing act that requires intense focus and an impressive array of gadgetry.

Standing 87 meters over Heathrow, the tower offers impressive views stretching miles over flat terrain. Even on the cloudy day of CNN's visit it's possible to pick out the tower blocks of London's financial district, nearly 30 kilometers to the east, and Windsor Castle to the west.

That incredible view—what Cooper calls "the greatest office window" in the region—is an essential part of the job. The controllers need visual confirmation of every takeoff and landing, only resorting to radar, radio and computers when the weather closes in.

Inside the tower there are two tiers—a raised central dais where two north and south runway controllers sit back-to-back, and a lower ring occupied by the ground controllers who govern aircraft from start-up at the stand until they've taxied into position on the runway and vice versa.

(Air traffic control company NATS has a cool 360-degree panorama of the tower on its website: http://bit.ly/1Hzo7N7)

By law the controllers can only work in 90-minute stretches punctuated by half-hour breaks over a six to nine-hour shift. These relatively short work stints help them focus on the task of moving as many planes as quickly as possible—up to 42 takeoffs per hour on a good day.

Mental juggling

Movement of planes on and off the ground is tracked using computerized strips containing flight data[Figure 2].

Figure 2　Computerized strip

Fingers dance over the touch screens as the controllers move around computerized strips (these used to be paper strips) that represent the planes and talk to pilots via the radio.

There's tricky mental juggling to perform as they figure out how to maximize takeoff slots so larger planes aren't followed by smaller aircraft that must wait longer for turbulence to clear.

At night there are extra complications guiding aircraft along Heathrow's busy taxiways. This is done using a giant touch screen that ATCs swipe to illuminate different stretches of Tarmac—possibly one of the world's largest lighting control decks.

There are also tracking monitors that show the position and movement of all aircraft on the ground, and screens detailing meteorological elements such as wind direction and speed.

Cooper shows off a relatively new piece of kit called the TBS—or time-based separation—that's proving crucial to Heathrow's traffic flow. Instead of spacing aircraft on approach by distance, it factors in wind speed and works out how many minutes apart they can land safely.

All good when everything runs smoothly, but what happens when it doesn't?

Much of the control tower's calm atmosphere seems to be down to the contingency measures in place for when problems arise.

It's typically overstaffed in case sudden sickness depletes numbers.

And says Cooper, almost every system is backed up at least two or three times.

Airspace scramble

Everything from the phones to the power supply is backed up two or three times in the tower [Figure 3].

Figure 3　Baclked up System

In the event of a power cut, the tower has an emergency generator. If that fails there are batteries. And if those go, staff can quickly relocate to a standby tower facility in a secret location nearby.

The controllers really earn their money (in UK more than MYM150,000 per year including overtime for senior ATCs) when disaster strikes and their cool nerves are tested to the limit.

Touch screen telephones in the control tower feature an oversized "crash" button, which instantly puts the ATCs in immediate contact with the emergency services and ground crews they need to coordinate with during major incidents.

Cooper recalls a 2013 incident involving a fire on board an empty, parked Ethiopian Airlines Boeing Dreamliner that closed the airport and led to a scramble to clear nearby airspace.

"It was a very intense day at the office," he says. "Most days here are hard work, some are harder than others. You don't get a job as a Heathrow controller and expect to put your feet up for a couple of hours."

Few people, it seems, are cut out for air traffic controlling.

The qualifications required are minimal, but ATCs need an innate range of skills that allow them to perform the mental acrobatics needed to keep everything flowing. (Try these games on the NATS website: http://bit.ly/1GyKaTa)

Only one in every 300 applicants makes the cut but they come from all walks of life. Cooper says Heathrow employs ATCs in their 20s barely out of school alongside staff with multiple doctorates to their name.

The training is necessarily lengthy, taking up to two years.

Even when qualified, ATCs can spend 18 months working under guidance before being allowed to operate solo. The same goes for experienced controllers moving to a new airport.

And it never stops.

Coolest "Xbox" ever

Heathrow's control tower simulator is equipped with all the same gadgets as the real thing [Figure 4].

Cooper takes us to a non-descript office block a few miles away from Heathrow where an almost full-sized control tower simulator has been constructed to help train and prepare ATCs and allow them to road test new aircraft and airport structural changes.

It's like a gigantic, awesome MYM2 million Xbox console with a 360-degree screen.

"We use it to put our ATCs in nasty situations," says Cooper, his fingers instinctively tapping the screen to line computer-generated aircraft up on the runway. "But it has to be relevant, you can't just come in here and set everything on fire just for the hell of it."

Figure 4   Tower Simulator

Sim manager Daniel Johnson puts his virtual tower through its paces, showing us weather conditions from thick fog to driving snow. He then takes us on a weird, stomach-flipping, out-of-body experience, sending the tower whizzing through the airport and out over London.

We also get to try our hand at clearing a few virtual jets for takeoff.

Even though it's not real, talking to the pretend pilot (actually Johnson in the next room) and moving the aircraft strips on the computer to get a pixilated Virgin Atlantic 787 off the ground proves a nerve-fraying experience.

Cooper isn't too impressed with the yawning gaps we allow between takeoffs, but offers some consolation.

"Even for us every day is a school day, you learn something about how to do the job differently.

"The place changes every day and you have to stay on top of the changes, but I cannot imagine having a job where you didn't."

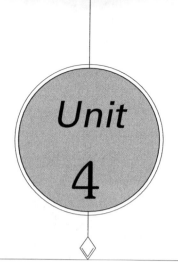

Unit 4

Aircraft Weight and Balance

I Reading Text

Aircraft Weight and Balance Definitions and Scope
in Maximum Takeoff Weight (MTOW)

The reference datum is an imaginary vertical plane from which horizontal distances are measured for aircraft weight and balance purposes. The reference datum is at location "0" and measurements for other reference points, like the baggage area or the passenger seats, are made in relation to the reference datum. The datum is determined by the manufacturer, and in small aircraft, the reference datum is often located along the firewall or at the leading edge of the wing.

Station

In terms of weight and balance on an airplane, the station is a location along the airplane fuselage given in terms of distance from the reference datum.

Arm

The arm is the horizontal distance from the reference datum to the center of gravity (CG) of an item.

CG Arm

The CG arm (where CG stands for center of gravity) is the arm obtained by adding the aircraft's individual moments and dividing the sum by the total weight of the unit.

Moment

A moment is the product of the weight of an item multiplied by its arm. (Moment/1,000 is used to simplify digits in some cases).

Center of Gravity (CG)

And aircraft's center of gravity is the point at which it would balance if it were suspended in air. Its distance from the datum is found by dividing the total moment by the total weight of the airplane. The center of gravity can be thought of as where all of the aircraft's mass is concentrated, or the "heaviest" part of the airplane.

Center of Lift

The center of lift is the point along the chord line of an aircraft wing or airfoil at which the force of lift is concentrated.

CG Limits

The forward and aft center of gravity locations for which the airplane must be operated within are referred to as CG limits. CG limits are based on a given weight.

Chord

The chord or chord line, of a wing, is an imaginary line representing a straight-line distance from the leading edge to the trailing edge of an airfoil.

Standard Empty Weight

The empty weight of an aircraft is the weight of the aircraft without including passengers, baggage, or fuel. Standard empty weight usually includes unusable fuel, full operating fluids, and full engine oil.

Basic Empty Weight

Basic empty weight of an airplane is the standard empty weight of the airplane plus optional equipment installed.

Maximum Landing Weight

The maximum landing weight is, as you might imagine, the maximum aircraft weight limit approved for an aircraft to land. Landing above this weight can cause structural damage.

Maximum Ramp Weight

The maximum weight for maneuvering on the ground is called the max ramp weight. Max ramp weight includes the weight of fuel used for start, taxi, and aircraft run-up procedures.

Maximum Takeoff Weight

The maximum weight limits for an aircraft to begin its takeoff roll is called the max takeoff weight.

Useful Load

The useful load is the difference between ramp weight or max allowable weight and basic empty weight. Useful load is the weight of the useful items on board, such as passengers and baggage.

Payload

An aircraft's cargo, baggage, and passengers (including pilots) make up its payload.

Load Factor

The ratio of the amount of load and aircraft that can withstand its maximum weight is called the load factor.

Tare

Tare is the weight of chocks, blocks, stands, etc. used when weighing an airplane. Tare weight is included in the scale readings and deducted from the scale reading to obtain the actual (net) airplane weight.

Standard Weights of Fluids

- Fuel: 6 lbs/gal
- Oil: 7.5 lbs/gal
- Water: 8.35 lbs/gal

In the FAA's Handbook of Aeronautical Knowledge, Max Takeoff Weight is defined as the "maximum allowable weight for takeoff." More specifically, the maximum takeoff weight is a limitation placed on the aircraft by the aircraft manufacturer during the design and testing process. It's a fixed weight.

Industry Jargon

In aviation, MTOW is short for Maximum Takeoff Weight. Sometimes people refer to this weight as MGTOW or Maximum Gross Takeoff Weight.

These two terms are interchangeable. More rarely, this weight might be referred to Maximum Brake Release Weight.

Maximum Takeoff Weight or Maximum Gross Takeoff Weight should not be confused with the Maximum Gross Weight of the aircraft itself, which is the maximum weight an aircraft can structurally handle, whether taking off or sitting on the ramp. The Maximum Gross Weight for an aircraft should never be exceeded at any time. MTOW can be sometimes exceeded, but not usually by a significant amount. For example, an aircraft can exceed MTOW when sitting on the ramp but must get rid of this weight before its brakes are released on the departure runway. Since the aircraft will burn fuel during the startup and taxi process, it's possible that it will weigh slightly more upon startup than during takeoff.

Importance

Due to structural limitations, an aircraft is restricted to a certain weight while taking off.

If this weight is exceeded, the aircraft may become structurally damaged, or worse, fail to complete a successful takeoff altogether.

Aircraft designers and manufacturers know the importance of MTOW. To an aircraft manufacturer's customer, a higher MTOW means an airplane can take off with more fuel and will have a longer range.

Use Caution

It's important to note that just because an aircraft is certified for a particular Maximum Takeoff Weight, it doesn't mean that the aircraft can always take off at this maximum weight. Many individual factors are to be considered for an aircraft to be determined safe to take off at a particular weight. A pilot needs to compute takeoff and climb performance, which is largely dependent on other variables like the following:

- Elevation: The higher the field elevation, the thinner the air is. An airplane will have a decrease in performance at high altitudes, which means a full payload may not be possible.
- Temperature: High temperatures also decrease aircraft performance, and can require a lighter load.
- Density Altitude: The higher the density altitude, which is pressure altitude corrected for nonstandard temperature, the worse the aircraft performance is.
- Runway length and surface: An aircraft loaded to its maximum takeoff weight might require a long runway, and the same aircraft may not be able to take off on a shorter runway under certain condition.
- Runway gradient: An up-sloping runway will require a longer takeoff distance than a down-sloping or flat runway, which should be taken into consideration with a heavy aircraft.
- Wind strength: A headwind aids takeoff performance; a tailwind degrades it.
- Obstacles during departure: Extremely heavy aircraft won't have a very good climb rate; therefore, it's especially important to calculate the climb gradient and rate for aircraft at

the particular takeoff weight. A required climb gradient to overcome obstacles may not be possible in a very heavy aircraft.

Alternate Spellings.

Max Gross Takeoff Weight, Max Takeoff Weight.

II   Words and Expressions

datum	*n.* 数据
firewall	*n.* 防火墙,隔火墙
fuselage	*n.* 机身
gravity	*n.* 重力,引力
moment	*n.* 矩,力矩;片刻,瞬间;因素,要素
suspended	*adj.* 悬浮的,暂停的,缓期的
chord	*n.* 和弦,和音
unusable	*adj.* 不能使用的
ramp	*n.* 支路,岔道,匝道
tare	*n.* 包装重量,皮重;车辆皮重,车身自重
deduct	*vt.* 减去,扣除
jargon	*n.* 行话,术语
startup	*n.* 启动所需的资金;新创办的小公司
exceeded	*adj.* 非常的;过度的;溢出的
ramp	*n.* 登机舷梯
taxi	*vi.* 滑行
elevation	*n.* 海拔
gradient	*n.* 斜坡,坡度
headwind	*n.* 逆风,顶头风
tailwind	*n.* 顺风

in terms of	根据,依据,按照,用
run-up	前夕

Ⅲ Exercises

1 Translate the following phrases into English

(1)最大起飞重量

(2)结构损伤

(3)最大松刹车重量

(4)结构限制

(5)最大着陆重量

(6)参考数据

(7)跑道梯度

(8)最大允许起飞重量

(9)航空知识手册

(10)爬升性能

2 Translate the following sentences into Chinese

(1)Extremely heavy aircraft won't have a very good climb rate; therefore, it's especially important to calculate the climb gradient and rate for aircraft at the particular takeoff weight.

(2)An airplane will have a decrease in performance at high altitudes, which means a full payload may not be possible.

(3)The empty weight of an aircraft is the weight of the aircraft without including passengers, baggage, or fuel.

(4)It's important to note that just because an aircraft is certified for a particular Maximum

Takeoff Weight, it doesn't mean that the aircraft can always take off at this maximum weight.

(5) The higher the density altitude, which is pressure altitude corrected for nonstandard temperature, the worse the aircraft performance is.

IV Supplement Reading

How Real is the Hypersonic Aircraft Revoluation?

By Miquel Ros

We live in an era of fast technological change: self-driving cars, drones, artificial intelligence.

Yet the tube-shaped subsonic airliners we keep flying on wouldn't look out of place in the 1960s.

Take the Boeing 737 for example.

A 50-year-old design remains one of the workhorses of the airline industry.

And going strong: Its latest iteration, the Boeing 737 MAX is expected to enter service next year.

To be fair, although from the outside it may look structurally similar to its earlier versions, decades of cumulative improvements have made the airliner of today a vastly more sophisticated, efficient and reliable machine.

Aircraft-making is an extremely capital-intensive activity and, given the financial and technical risks that launching an entirely new model entails, it's understandable that the industry prefers to keep milking proven concepts.

However, how long before the current generation of airliners reaches its limits?

From electric propulsion to hypersonics, from NASA to private entrepreneurs, the quest for new, truly groundbreaking, aircraft concepts are on.

And it has the potential to forever change our idea of air travel.

High-voltage innovation

Airbus, for example, has unveiled its future aircraft concept.

This isn't exactly a new aircraft program, but a depiction of what would be possible if all of the futuristic technologies envisaged by Airbus could be combined to create the ideal airliner.

Across the Atlantic, Boeing is also working, together with NASA, on a number of futurist aircraft concepts within the framework of the New Aviation Horizons initiative.

More: Why is the sun setting on the Boeing 747?

The SUGAR program (that stands for Subsonic Ultra Green Aircraft Research) has come up with some truly innovative aerodynamic and propulsion solutions.

These include an aircraft with eye-catching truss-braced wings and an hybrid gas-electric propulsion system fed by liquefied natural gas.

The search for new modes of propulsion is particularly important, as aviation remains one of the few major industries where replacing fossil fuels remains an unresolved challenge.

Biofuels may offer a stopgap solution, as they can be adapted to fit current engine technology and supply infrastructure. However, it's electrically powered flight that's captured the attention of a handful of visionaries.

It's a technology still in its infancy, but one that benefits from the enthusiasm and resourcefulness of entrepreneurs, not unlike the mavericks of the early days of aviation.

By competing with each other to break the next record, they contribute to the advancement of the aeronautical science.

More: Space tech meets aviation

In 2015, as the long-winged Solar Impulse tried to circumnavigate the globe on solar power, teams were vying to be the first to cross the English Channel on an electric-powered aircraft [Figure 1]. French scientists and former yachtsman Raphael Dinelli are also preparing a solo crossing of the Atlantic later this year on a hybrid biofuel-electric light aircraft called Eraole.

His plane derives part of its energy from solar power. If successful, a derivative of Eraole might soon be serially produced for the private aviation market.

Figure 1　Solar impulse plane

Beyond the boom

Electric and hybrid aircraft will make flying greener, but what about speed?

Significantly increasing the speed of current jetliners means, inevitably, breaking the sound barrier, which presents a whole set of challenges, not all of them technological.

Supersonic flight isn't exactly new: The Concorde linked both sides of the Atlantic for more than three decades until economic and political issues led to its retirement in 2003.

The sleek Franco-British airliner remained an aeronautical curiosity, an experiment without continuity or replacement.

NASA has been particularly active in developing a new generation of more efficient, quieter supersonic airliners to revive commercial supersonic air travel as a viable proposition.

It worked on such concepts with Boeing through the New Aviation Horizons initiative, and recently teamed up with Lockheed to research Quiet Supersonic Technology (QueSST).

The key is to find a way to smother the sonic boom that's produced whenever an aircraft breaks the sound barrier.

Concorde, for example, was only allowed to make full use of its supersonic capabilities when flying over the ocean, thus significantly limiting the number of markets it could serve.

A lower sonic boom may allow a future supersonic airliner to fly routes over land, vastly increasing potential markets.

Speed will still come at a cost, though. This is why any supersonic comeback is likely to start with those that are most able to pay for it.

The Aerion Corporation, a Nevada-based private aircraft manufacturer, and Airbus have already started work on a supersonic private jet, the Aerion AS2[Figure 2].

Expected to enter service early in the next decade, it'll be able to carry up to 12 passengers at speeds of Mach 1.6.

Figure 2   Supersonic Aerion AS2

Hypersonic hype

Once you've broken through the sound barrier, why not double down?

Although still closer to sci-fi than the tangible realities of today's aviation industry, several research organizations, from Europe to Japan, are making inroads into hypersonic flight.

We're talking about aircraft capable of Mach 5 to 8, five to eight times faster than sound.

Realistically these revolutionary concept aircraft—with names like Lapcat and the Hikari—are several decades away, but they're starting to appear like a very feasible possibility.

More: Hypersonic aircraft goes by 5 times the speed of sound

One of the most ambitious concepts in the field of hypersonics is the SpaceLiner, being developed at the German Aerospace Center (DLR).

The SpaceLiner applies space technology to commercial aviation in order to achieve speeds of up to Mach 25, enough to travel from London to Australia in under 90 minutes.

In achieving this amazing speed, the SpaceLiner takes its passengers to the edge of space [Figure 3].

In fact, it's a two-stage concept, reminiscent of the, now retired, space-shuttle.

A booster takes a civilian-carrying stage to a height of roughly 80 kilometers, where the passenger vehicle detaches itself to carry up to 50 passengers to the other side of the globe.

Figure 3　Spaceliner

In line with stringent safety requirements, the passenger cabin can also double as a rescue capsule.

Both the booster and the capsule are fully reusable, an essential requirement to keep costs under control.

That's a principle well understood also by SpaceX and the emerging private space industry, that has also focused on developing reusable space vehicles.

Dr. Olga Trivailo, a researcher at the German Space Center, DLR, says the SpaceLiner is also an environmentally friendly concept, using a rocket propellant mix of liquid hydrogen and liquid oxygen that produces only water vapour upon combustion.

Thus hydrogen has potential as a non-fossil fuel alternative, although a way would first need to be found to manage the associated higher costs compared with kerosene, predominantly due to new infrastructure requirements.

Dr. Martin Sippel, head of the Space Launcher Systems Analysis department at DLR's Institute of Space Systems in Bremen, says it would be reasonable to expect wider use and implementation of hydrogen propellants in the next 35 to 50 years.

The bionic cabin

While it's difficult to anticipate which one of these different approaches to the aircraft of the future will prevail, one thing seems sure: The air travel experience will be transformed.

Ergonomic and lighting improvements such as those found on the newest airliners today are just a foretaste of what lies ahead.

Related Content

The Antipode: Flying from New York to London in 11 minutes

Even if Airbus' vision of a "bionic smart cabin" made of natural, smart materials that adapt to the needs of each passenger is only partly realized, the scope for improvement is massive.

Paradoxically, because of aerodynamic requirements, the supersonic and hypersonic aircraft of the future may well be windowless.

The need to make up for the lack of windows, plus the seating solutions that it opens up, is likely to spur a new wave of innovation that'll further redefine the in-flight experience.

The aircraft of the future will certainly take us to our destination in a faster, greener and more comfortable way.

What's not guaranteed is that we'll still get our window seat.

Unit 5

Airport Delay

I Reading Text

World's Worst Airports Ranked by 2015 on-time-departure ratings

1. Jakarta, Indonesia	35.22%
2. Hangzhou, China	40.58%
3. Nanjing, China	44.65%
4. Shanghai-Pudong, China	51.62%
5. Shenzhen, China	53.93%

Danny Armstrong is used to deal with flight delays and cancellations. As general manager of China Banking at National Australia Bank, he travels for business around China a couple of times a month. At the mercy of the nation's notoriously unpredictable flight system, Armstrong had to become adept at contingency planning: booking on a certain carrier he believes has better on-time performance and switching to the high-speed train during times of the year when delays are common.

On occasion, he's taken even more drastic measures to be punctual, including the time when he hired a police escort from the airport to get the CEO of his company, on a visit from Australia, to a high-level meeting after flight delays from Shanghai to Beijing. "We just did it by the skin of our teeth," he said. "It cost us a fair bit of money, as well."

China's airports are the worst in the world when it comes to punctuality.

Armstrong's frustrations are common among business travellers in China as air travel has become increasingly chaotic and unreliable in recent years due to the country's chronically—and well-chronicled—overcrowded skies. Airport statistics website FlightStats last year ranked 188 midsized and large airports around the world according to how many flights departed on time.

Of the bottom 20, 14 of them were in China, Hong Kong or Taiwan—all of which had on-time ratings of less than 60%. The least punctual Chinese airport was Hangzhou's at an on-time rating of 41%, second only to Jakarta globally. Shanghai's Pudong Airport (52%), Hong Kong (59%) and Beijing (64%) fared only marginally better. By comparison, Tokyo's busiest airport, Haneda, which handles more flights annually than Shanghai-Pudong, had an on-time rating of 92% last year, among the best in the world.

The rankings were even worse in 2014. Just more than a third of flights took off on time at the four biggest airports in the Yangtze Delta region (Shanghai-Pudong, Shanghai-Hongqiao, Hangzhou and Nanjing)—making them the most delayed in the world by far.

Airspace in short supply

Aviation experts agree that one of the main problems in China is the fact that the country's airspace is largely controlled by the military, leaving little room for civilian aircraft even as the domestic airline industry booms. According to the state-run China Daily newspaper, less than 30% of China's airspace can be used by commercial airlines, compared to about 80% of the airspace in the United States.

Not only does this create a daily bottleneck for passenger planes, it can occasionally result in mass flight cancellations when the military wants to conduct drills, such as in the summer of 2014 when 12 airports, including the two in Shanghai, were ordered to reduce traffic by 25% for a three-week period for military training.

China is projected to overtake the United States as the number-one business travel market in the world by 2017.

And yet the airline industry continues to grow in China, with airlines buying more planes, selling more tickets and putting considerably more strain on an already beleaguered system. Chinese airlines were estimated to have carried 440 million passengers last year, an 11% increase over the year before. According to the International Air Transport Association, they're expected to add another 758 million new passengers over the next two decades, bringing the total number of annual passengers to 1. 2 billion by 2034.

Increased leisure travel is full much of this demand, but according to the Global Business Travel Association, business travel is also booming. The group estimates that business travellers in China spent MYM261 billon in 2014—an increase of 16. 6% over the previous year—and despite the slowing economy, it projects that China will overtake the US as the No. 1 business travel market in the world sometime this year.

To accommodate all these new flyers, the Chinese government is starting to make more room in the skies. Last October, authorities opened a third of the country's low-altitude airspace (below 1,000 m) to civilian aircraft, a move that will benefit mostly emergency services and helicopter tours. The plan is to eventually expand this to 3,000 m.

"The government realizes there's an airspace constraint issue," said David Yu, the executive director in Asia for the International Bureau of Aviation, a UK-based consultancy. "The question: 'Is air space opening up and is it going in that direction?' The answer is yes. The question: 'How fast?' That's a hard one to answer."

China is also aggressively building more airports and striving to improve the efficiency of its existing facilities by increasing the number of flights that can safely land and take off every hour, said Steven Brown, the chief operating officer of the Washington-based National Business Aviation Association, who has flown in China for 30 years.

To achieve that, he said, the country is adopting the best practice in the aviation industry, from training crews to be more efficient as they maneuver the planes to improving its air-traffic control capabilities and the lighting on the runways. It's only a matter of time, he added, before the country's infrastructure and systems are on par with the West. China is developing very, very quickly, far faster than any other country around the world has ever developed that capability.

Dead time in airport

For companies that rely heavily on air travel to do business in China, all these improvements can't happen fast enough. Lost productivity from chronic flight delays is perhaps the biggest complaint at the moment. "It's had a big impact on us," said Ray Chisnall, the Asia-Pacific director for Gleeds, a London-based construction consultancy that manages building projects for clients across China. "If our people are spending a lot of time travelling, they're either not earning fees or they're not doing work for clients. It's just dead time. "

Chisnall estimates his senior managers waste between 10% ~ 15% of their time "hanging around in airports".

With up to 30 ~ 35 employees travelling around the country to building sites each month, the company has been forced to rethink the way it operates. Daytrips to other cities for meetings are a thing of the past. "Well, you can do it if you accept you're going to get home at 2:00 or 3:00 in the morning—if you get home. " Chisnall said. His staff usually spend the night in a hotel and have dinner with a client instead. Everyone on his team has 4G-equipped phones so they can reply to emails when stuck for hours in airports with spotty wi-fi.

Chisnall doesn't keep track of the cost to the company in lost productivity. "We probably should to be honest," he said, "it would probably scare us if we did. "

The other cost to companies is in lost business or frayed relationships due to missed meetings. Armstrong said he's been forced to make excuses in the past when a flight's been cancelled or delayed, which he calls "dreadful," particularly when he's meeting with banking regulators in Beijing. "They have zero flexibility. They consider it an affront," he said. "If you are late for a meeting. . . that can potentially impact on future licensing aspirations you might have or a variety of different things that aren't immediately transparent to you. "

It's difficult to quantify the impact of flight delays and cancellations have had on the overall economy, but Li Xiaojin, a professor at the Civil Aviation University of China, told China Daily in 2013 that if the military transferred just 10% of its airspace to passenger planes, it could boost China's GDP by 200 billion RMB (MYM32. 6 billion).

Rage in the skies

It's one thing if delays are a rare occurrence, but in China, every business traveller has a nightmarish story to tell. Peter Arkell, managing director for the Australian HR consulting firm Carrington Day, said he once had a 19:00 flight out of Shanghai that was delayed more than five hours due to "air traffic control" issues, a frustratingly vague explanation often provided by airlines for delays. He boarded after midnight and the plane then sat on the tarmac for another three hours before the flight was cancelled without explanation and everyone was offloaded, bleary-eyed, and sent to a hotel. The plane finally took off at midday of the next day.

"When there are delays, there are big-time delays. The ones that annoy you are not the weather... It's the ubiquitous air-traffic control excuse," he said. "To be locked up in a small space like that and just not going anywhere and not having any information that can drive you really crazy." This is evident in the rash of embarrassing air-rage incidents in China in recent years, from passengers storming the runway to starting riots to opening the emergency doors during taxiing.

To reduce the stress of travel, many business people have stopped flying altogether on short routes and rely on the high-speed train network, which is punctual, comfortable and expanding at a remarkable clip. The Shanghai – Beijing route, for instance, takes about five hours, with trains departing every 10 to 30 minutes throughout the day. "There was a really bad period a year ago when I just took to the train and caught it for about four months," Armstrong said. "It's a far more pleasant experience and you can actually work all the way."

But as frustrating as air travel in China, it's not going to stand in the way of companies continuing to invest in the country—at least not yet. "I think a lot of people just factor it in," Chisnall said, "and accept it as one of those costs of doing business here."

Ⅱ Words and Expressions

mercy	*n*. 仁慈,宽容;怜悯;幸运
notoriously	*adv*. 众所周知地;声名狼藉地;恶名昭彰地
adept	*n*. 内行;能手
	adj. 熟练的;擅长……的
contingency	*n*. 偶然性;意外事故;可能性
drastic	*adj*. 激烈的;猛烈的
punctual	*adj*. 准时的,守时的;精确的
escort	*n*. 陪同;护航舰;护卫队;护送者
	vt. 护送;陪同;为……护航
frustration	*n*. 挫折
chaotic	*adj*. 混沌的;混乱的,无秩序的
chronically	*adv*. 长期地;慢性地;习惯性地

fare	*vi.* 经营;进展;遭遇;过活
	n. 票价;费用;旅客;食物
marginally	*adv.* 少量地;最低限度地;在栏外;在页边
Haneda	*n* 羽田(日本机场名)
boom	*vt.* 使兴旺;发隆隆声
	vi. 急速发展;发隆隆声
	n. 繁荣;吊杆;隆隆声
state-run	*adj.* 国营的;州立的
bottleneck	*n.* 瓶颈;障碍物
occasionally	*adv.* 偶尔;间或
drill	*n.* 训练;钻孔机;钻子;播种机
	vi. 钻孔;训练
	vt. 钻孔;训练
overtake	*vt.* 赶上;压倒;突然来袭
	vi. 超车
strain	*n.* 张力;拉紧;负担;扭伤
	vi. 拉紧;尽力
	vt. 拉紧;滥用;滤去;竭力
beleaguer	*vt.* 围攻;围
accommodate	*vt.* 容纳;使适应;供应;调解
	vi. 适应;调解
authority	*n.* 权威;权力;当局
helicopter	*n.* 直升机
consultancy	*n.* 咨询公司;顾问工作
aggressively	*adv.* 侵略地;攻击地;有闯劲地
striving	*n.* 努力;斗争
	v. 努力(strive 的现在分词);奋斗;力争
maneuver	*n.* 机动;演习;策略;调遣
	vi. 机动;演习;调遣;用计谋
	vt. 机动;演习;用计;调遣
par	*n.* 标准;票面价值;平均数量
	adj. 标准的;票面的

complaint	*n.* 抱怨;诉苦;疾病;委屈
stuck	*v.* 刺(stick 的过去式)
	adj. 被卡住的;不能动的
spotty	*adj.* 多斑点的;质量不一的
scare	*vt.* 惊吓;把……吓跑
	vi. 受惊
	n. 恐慌;惊吓;惊恐
fray	*n.* 争论;打架;磨损处
	vt. 使磨损;变得令人紧张、急躁
	vi. 被磨损
affront	*vt.* 冒犯,面对;公开侮辱
	n. 轻蔑;公开侮辱
aspiration	*n.* 渴望;抱负;送气;吸气
transparent	*adj.* 透明的;显然的;坦率的;易懂的
nightmarish	*adj.* 可怕的;噩梦似的;不愉快的
frustrating	*adj.* 令人沮丧的
	v. 使沮丧(frustrate 的现在分词)
vague	*adj.* 模糊的;含糊的;不明确的
tarmac	*n.* 柏油碎石路面;柏油碎石飞机跑道
offload	*vt.* 卸下;卸货
	adj. 卸载的
bleary	*adj.* 朦胧的;眼睛模糊的
annoy	*vt.* 骚扰;惹恼;打搅
	vi. 惹恼;令人讨厌;打搅
ubiquitous	*adj.* 普遍存在的;无所不在的
rash	*adj.* 轻率的;鲁莽的;不顾后果的
embarrassing	*adj.* 使人尴尬的;令人为难的
riot	*n.* 暴乱;放纵;蔓延
	vi. 骚乱;放荡
	vt. 浪费,挥霍
keep track of	记录;与……保持联系

III   Exercises

1   Translate the following phrases into English

(1) 国际航空运输协会

(2) 全球商务旅行协会

(3) 长江三角洲地区

(4) 航班延误和取消

(5) 空域限制问题

(6) 低空空域

(7) 在机场徘徊

(8) 本土航空工业

(9) 打开应急舱门

(10) 高铁运输网络

2   Translate the following sentences into Chinese

(1) Armstrong's frustrations are common among business travellers in China as air travel has become increasingly chaotic and unreliable in recent years due to the country's chronically—and well-chronicled—overcrowded skies.

(2) Not only does this create a daily bottleneck for passenger planes, it can occasionally result in mass flight cancellations when the military wants to conduct drills, such as in the summer of 2014 when 12 airports, including the two in Shanghai, were ordered to reduce traffic by 25% for a three-week period for military training.

(3) China is also aggressively building more airports and striving to improve the efficiency of its existing facilities by increasing the number of flights that can safely land and take off every hour, said Steven Brown, the chief operating officer of the Washington-based National Business

Aviation Association, who has flown in China for 30 years.

(4) Armstrong said he's been forced to make excuses in the past when a flight's been cancelled or delayed, which he calls "dreadful," particularly when he's meeting with banking regulators in Beijing.

(5) To achieve that, he said, the country is adopting the best practices in the aviation industry, from training crews to be more efficient as they maneuver the planes to improving its air-traffic control capabilities and the lighting on the runways.

IV   Supplement Reading

Floating airports: Could they finally become a reality?

By Miquel Ros

Airports need a lot of space, but in the places most in need of air connections—islands and large metropolises—it's a commodity that's in short supply.

In search of solutions, planners will sometimes turn to the sea.

Huge land reclamation projects are an obvious route. That's what was done in Hong Kong and Osaka to expand airport capacity.

Some visionaries have taken a rather different approach.

What if we could make a runway float over the virtually limitless flat surface of the sea?

Floating deck

Aircraft carriers have floating runways, but they're also warships which need to be maneuverable and to travel at speed.

Their decks are too small to deal with commercial craft—even the largest carrier falls short of the requirements of modern jet airliners.

However, let's take the concept of the floating deck further.

What if we strip it of its engine and lower deck, anchor it in a set location, and then make it long enough and wide enough for a medium-sized airliner to land on it?

The result is a very large floating structure that's neither a ship nor an island: It's a floating airport.

Forerunner from history

During World War II, the British entertained the idea of building runways on icebergs in order to provide cover for the vital Atlantic convoys.

Project Habakkuk never materialized, but the floating airport concept lived on.

In 1995, 17 Japanese private firms, mostly shipbuilders and steelmakers, formed the Technological Research Association of Mega-Float with the support of the Japanese government.

The goal was to design and test a floating airport concept that, if successful, could be installed in Tokyo Bay and the Tokyo Mega-Float is possibly the most ambitious attempt of its kind to date.

The project called for a floating structure with a 4,000-meter-long runway, enough even for large airliners.

A smaller-scale model with a 1,000-meter-long runway was actually built and tests proved that the Mega-Float was suitable for aircraft operations.

However, the project did not go ahead and the structure was later dismantled.

San Diego proposal

Similar proposals have been floated, quite literally in this case, for San Diego, a city whose international airport has little room left to expand in its current location.

Proposals to build a brand new two-runway international airport in the sea, a few miles off Point Loma have been put forward by two different companies, OceanWorks Development and Float Inc.

Whether it was the MYM20 billion price tag that proved to be a bit too expensive, or lingering doubts about the technical feasibility of the concept, the fact is that neither of the project has been materialized.

Floating aerotropolis

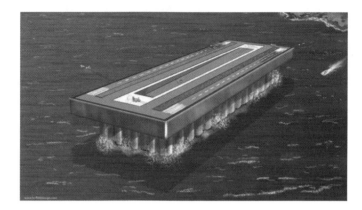

Figure 1   An artist's rendering of Terry Drinkard's floating airport concept

The San Diego projects have some similarities with the floating airport concept devised by Terry Drinkard, an American aeronautical engineer who has conducted extensive research in this field[Figure 1].

His scheme draws heavily from technologies and materials that have already been tested in the construction of deepwater oil rigs.

Drinkard's vision is for a full-fledged offshore "aerotropolis": a floating structure, as well as being able to handle medium-sized airliners (of the Boeing 737 or A320 types), would also host a whole range of economic and research activities, from experimentation with renewable energy technology to aquaculture and yachting.

This floating airport would be energy self-sufficient.

Power would be harvested from the wave, from the sun and through ocean thermal energy conversion, a technology that produces electricity by utilizing the temperature differences between depths of seawater.

Its structure would provide a base for oceanographic research or aquaculture and it would also double as a port and recreational marina, while its potential offshore status could attract a host of other economic activities.

Bridging the Atlantic

A more daring version of the concept revives the idea, already put forward by a 1930 article in the magazine Popular Mechanics, of bridging the Atlantic with a chain of such floating airports.

This would enable all sorts of aircraft, including smaller airliners and private jets, to cross the ocean without the need to secure costly ETOPS (extended-range twin-engine operational standards) certification or take the more indirect northern route through Iceland and Newfoundland.

However, the most realistic chances for a floating airport might be found in warmer water.

Commander Bud Slabbaert, aviation consultant and Drinkard's partner in this project, says a small-scale version of the floating airport is currently being evaluated by several governments and infrastructure operators in the Caribbean.

It's a region heavily dependent on air transport where the lack of space is compounded by some of the islands' hilly or mountainous terrain.

Other aviation experts remain a lot more skeptical when consulted about the feasibility of the concept.

Mann R W Mann, a consultant and former airline planning executive with several U. S. network airlines, thinks such an project is highly unlikely to get off the ground due to the relatively high cost and limited practical gain.

"Projects that involve building runways on the sea, such as those currently underway in the South China Sea, are driven primarily by non-economic considerations. " he says.

Boris Island

It's true that airports are a massive investment—even more so if they have to be built over water.

So it's no surprise that, despite the concept of a floating airport having being around for a long time, it's failed to come to fruition yet.

Despite decades of increasing air traffic and sprawling urban development across the world, airport relocation is still a rarity—Hong Kong did it in 1998 and Munich did it in 1992.

In London, the controversy about the future of Heathrow airport, one of the most congested major air hubs anywhere in the world, has been raging on for years.

Some of the most remarkable proposals to sort out London's air capacity problem involve relocation to the Thames Estuary, where the River Thames meets the North Sea.

Renowned architect Norman Foster came up with the concept of a four-runway airport to be built from scratch on the Isle of Grain, a marshy area some 30 miles east of central London.

The project was popularly known as "Boris Island," because former London mayor Boris Johnson was one of its most solid backers, but it was rejected by the UK Airports Commission in 2014.

Six runways for MYM63 billion

Figure 2   Gensler's proposal for southeast England's airport expansion

A rather more ambitious proposal was put forward by the architecture firm Gensler and Thames Estuary Research and Development (Testrad), calling for the airport to be built right in the middle of the estuary[Figure 2].

In this vision, a six-runway airport would be built over the water of the Thames Estuary at a cost of nearly MYM63 billion.

The runways would stretch parallel to each other on both sides of a central core, where the main terminal would be located.

The airport would be connected to land through underwater tunnel for high-speed rail connection.

Gensler considered two options, explained by Ian Mulcahey, the firm's managing director.

The first was a floating structure along the lines of the Tokyo Mega-Float, but the water in that area proved to be too shallow.

So it opted for a "polder"-like concept, where dykes would be used to delimit the area of the airport and water would then be drained to create a dry, flat surface.

"Although it might sound challenging when compared to building on dry land, by building the airport in a body of water we can get away with the lengthy and costly process of gathering the huge expanse of land you need for a project of this size," says Mulcahey.

He insists on the final decision about London airport expansion has which not yet been taken.

On June 30, the UK's Transport Secretary Patrick McLoughlin announced that following the UK voting to leave the European Union, such decision-making has been delayed until "at least October."

So while an extra runway at Heathrow and Gatwick is by far the most popular solution being debated, don't write off the Thames estuary gateway just yet.

A question of economics

What decades of studies about the feasibility of the floating airport concept has shown is that its ultimate development and application is not a matter of technology, but getting the economics right.

The oil industry has demonstrated that when there is a pressing need—and money to be made—virtually any obstacle can be pushed aside.

The vast number of oil platforms defying the element in some of the world's harshest environment are witness to it.

The same could be one day true of the floating airport.

Unit 6

Airport Collaborative Decision Making

I   Reading Text

The Silo Effect-Slowing Down Operational Efficiency at Airport

Air navigation service providers, airlines and airport operators know that challenges in the aviation industry are highly demanding and simply maintaining and sustaining day-to-day activities is becoming increasingly more difficult due to increasing traffic volume, heightened security concerns, and emerging technologies.

Under financial and environmental constraints, there is a need to maximize asset within current budget and with existing infrastructure. Stakeholders are tasked to attain maximum performance and service from their assets, while simultaneously delivering improved capacity and cost efficiencies.

Nevertheless, airport stakeholders often operate independent systems in isolation, focusing on their own outcomes and without a shared situational awareness across the wider airport community. This limited perspective on the operation as a whole can result in widespread inefficiencies.

Phil Ensor, a corporate director of Goodyear Tire, described in The Functional Silo Syndrome (1988), the organizational structure where lines of business are divided and isolated from one another leading to disfunction. He coined the now famous phrase "the Functional Silo

Syndrome" to describe Goodyear's organizational difficulties due to employees working in lack of communication. "…communication is heavily top-down-on the vertical axis. Little is shared on the horizontal axis, partly because each function develops its own special language and set of buzzwords".

Lack of Common Vocabulary and Definitions-Groups with limited interaction often develop their own semantic references; this includes airport stakeholders as they may use different terminologies to cover the same reality. This lack of common definition and understanding of terms and processes across the stakeholder community can exacerbate misunderstanding and contribute to the lack of common situational awareness.

As an example, "arrival time" to an air traffic controller (ATC) or an airline could mean at the point of touch down, whereas for ground handling agencies "arrival time" may be understood as the time when an aircraft is at the gate. This disparity in a common definition of terms leads to a lack of shared awareness and clarity of the operational picture, which can lead to confusion and result in increased inefficiencies.

Lack of Information Exchange and Communication-Another major factor that causes inefficiencies and misunderstanding among stakeholders is a lack of clear and concise communication and information exchange processes. There are many instances where departments and organizations work independently of each other and do not share information, data and concerns, which leads to decisions and actions being reactive rather than proactive and are based on incomplete or faulty information.

There are often no recognized processes or agreements bringing all stakeholders together regularly. This denies stakeholders an opportunity to share their difficulties and successes with other members of the airport community. In addition, a lack of process makes it difficult to ensure that relevant information is shared with concerned stakeholders in a timely manner.

A common theme when looking at challenges faced by an airport community is that stakeholders are simply not aware of the needs, constraints, and others' goals, and how decisions and actions impact on the other stakeholders and the operation of the system as a whole, often in an escalating fashion. The data and information required to enhance efficiency exist, and it must be shared within the network in real-time to facilitate better decision-making across the stakeholder community. This allows business partners to act more predictably-even during unusual situation, which is essential for a smooth overall process.

Disconnected Strategies and Working in Isolation-In an effort to improve their own performance, stakeholders may work independently to deliver increased efficiencies in their area without realizing the impact on other stakeholders and thus the overall operation, includes their own. These disconnected or fragmented procedures, strategies, and systems often lead to decreased efficiency and performance across the entire operation.

For example, slippery runway condition caused by snow may reduce runway capacity and create other constraints, such as aircraft departure being delayed. These delays may mean that more passengers are accommodated inside the terminal increasing pressure on existing check-in and security areas. In addition, if aircraft do not clear their gates in a timely manner, ground handling and other resources within the terminal are effected, potentially causing delays to inbound aircraft.

The entire airport community feels the impact of delays. Without shared situational awareness across all stakeholders, and the ability to appropriately plan, both tactically and strategically, the delays can compound and escalate, further deteriorating the situation and leading to a blaming culture among stakeholders.

Principles of Collaborative Decision-Making

The principle of CDM is to put in place agreed cross-collaborative processes including communication protocols, training, procedures, tools, regular meetings and information sharing, which moves ATM operations from stovepipe decision-making into a collaborative management process that improves overall system performance and benefits the individual stakeholders.

There are clearly defined Airport-CDM concepts and activities such as A-CDM in Europe, Surface-CDM in the USA, and CARATS in Japan. While there are clearly defined and shared Airport-CDM concepts and activities at the global level, the approach to the concepts and its implementation may vary at the regional level such as A-CDM in Europe, or at the ANSP level, such as Surface CDM in the USA or CARATS in Japan. However, in each case they share common objectives and strategies.

This document aligns CDM into the following categories and their respective area of focus:

En-route CDM focuses on routing and air traffic flow management operations

Airport-CDM/Airside A-CDM focuses on turnaround and surface management operations

Landside A-CDM focuses on operations inside the terminal

Airport-CDM is a process that applies to all airports irrespective of size that supports landside, airside, and en route air traffic flow management (ATFM) operations, while enhancing forward planning and tactical decision-making. ICAO has included A-CDM as one of the Aviation System Block Upgrades (ASBU) Block 0 Modules to improve airport operations. The related key performance areas (KPA) are capacity, cost, efficiency and environment.

IATA regards A-CDM as an "⋯ effective management of existing airport capacity" and mandates its Airport Working Group to "implement operational improvements at airports, such as Airport-Collaborative Decision Making (A-CDM)".

Olivier Jankovec, Director General of ACI Europe, stated in 2014 that "There is little doubt that A-CDM is the way forward and that it is a win-win-win-win for airports, airlines, ANSPs and the travelling public".

Sharing operationally relevant information among all stakeholders involved in airside processes, from the approach, to the turnaround and the take-off, is fundamental in achieving common situational awareness and a collaborative decision-making processes involving multiple stakeholders. By exchanging real-time pertinent information, stakeholders will share a common view that is based on the same and best information available, which will have a positive impact on predictability, punctuality, fuel burn, and environment as well as the utilization of resources. While A-CDM delivers many benefits on its own, to further enhance the benefits, sharing information between en-route and landside is essential.

Examples of airside Airport-CDM processes

Sharing essential time information—An aircraft has an estimated take—off time at 12:00 UTC and to achieve an on-time departure, the estimated arrival time of the in-bound leg, the aircraft ready time, and the taxi-time are needed. Even in this simple example, to have a punctual departure, collaboration and sharing information is needed between en-route, landside and airside operations.

Minimizing taxi queues when having high departure demand—When departure demand is greater than capacity, long departure queues can result, with aircraft burning fuel unnecessarily while waiting to depart. In this scenario, proactive A-CDM processes and sharing information can improve the situation by sharing information such as scheduled and estimated off-block and takeoff times as well as estimated taxi-out times. Processes can be put in place that enable the ANSP, airlines, and ground handlers to predict when to start up and push back from gate to minimize queuing at the runway.

Malfunction causing delayed departure-If boarding of a flight will commence 15 minutes later than planned due to a malfunctioning computer at the gate, this information will be available simultaneously to the ramp agent, gate planning, tow truck dispatcher, en-route and terminal ATCO, and possibly to entities at the affected flight's destination airport.

Runway closure due to snow removal-through A-CDM the closure of a runway due to snow removal will be known to involved stakeholders. This will allow new optimal start-up sequences to be calculated that take into account arriving aircraft, individual taxi times, de-icing requests and capacity restrictions of the surrounding airspace. This will also allow the optimization of available resources and allow aircraft to reach the runway holding points as soon as snow removal is complete and be able to depart without additional and unnecessary delay.

II Words and Expressions

asset	*n.* 资产;优点;有用的东西;有利条件;财产
navigation	*n.* 导航
constraint	*n.* 约束;局促,态度不自然;强制
infrastructure	*n.* 基础设施
stakeholder	*n.* 利益相关者;赌金保管者
simultaneously	*adv.* 同时
isolation	*n.* 隔离;孤立;绝缘;离析
awareness	*n.* 意识,认识;明白,知道
perspective	*n.* 观点;远景;透视图
	adj. 透视的
widespread	*adj.* 普遍的,广泛的;分布广的

syndrome	*n.* 综合症状；并发症状；校验子
organizational	*adj.* 组织的；编制的
disfunction	*n.* 功能紊乱；机能障碍；官能不良
	vi. 功能失调；出现机能障碍；垮掉
buzzword	*n.* 流行词
semantic	*adj.* 语义的；语义学的（同 semantical）
exacerbate	*vt.* 使加剧；使恶化；激怒
whereas	*conj.* 然而；鉴于；反之
disparity	*n.* 不同；不一致；不等
clarity	*n.* 清楚，明晰；透明
concise	*adj.* 简明的，简洁的
proactive	*adj.* 前摄的（前一活动中的因素对后一活动造成影响的）
manner	*n.* 方式；习惯；种类；规矩；风俗
escalating	*v.* 逐步上升（escalate 的现在分词）
	adj. 逐步上涨的
enhance	*vt.* 提高；加强；增加
facilitate	*vt.* 促进；帮助；使容易
fragmented	*adj.* 成碎片的
	vt. 分裂（fragment 的过去分词）；使成碎片
slippery	*adj.* 滑的；狡猾的；不稳定的
tactically	*adv.* 战术性地；策略高明地
strategically	*adv.* 战略性地；战略上
deteriorating	*v.* 退化，恶化（deteriorate 的现在分词）
blaming	*vbl.* 归咎；责备；谴责
protocol	*n.* 协议；草案；礼仪
	vt. 拟定
	vi. 拟定
stovepipe	*n.* 大礼帽；火炉烟囱
	adj. 瘦裤腿的
turnaround	*n.* 转变；转向；突然好转；回车道
irrespective	*adj.* 无关的；不考虑的；不顾的

modules	n. 单元
mandate	n. 授权;命令,指令;委托管理;受命进行的工作
	vt. 授权;托管
pertinent	adj. 相关的,相干的;中肯的;切题的
utilization	n. 利用,使用
queue	n. 队列;行列
	v. 排队
scenario	n. 方案;情节;剧本;设想
ramp	n. 斜坡,坡道;敲诈
	vi. 蔓延;狂跳乱撞;敲诈
	vt. 敲诈;使有斜面
tow	n. 拖;麻的粗纤维;拖曳所用之绳
	vt. 拖;牵引;曳
	vi. 被拖带;拖行
dispatcher	n. 调度员;调度程序;分配器
destination	n. 目的地,终点
de-ice	adj. 除冰的
	vt. 除冰;除冰带
optimization	n. 最佳化,最优化
day-to-day	adj. 日常的;逐日的
real-time	adj. 实时的;接到指示立即执行的
start-up	n. 启动
	adj. 起动阶段的;开始阶段的
top-down	adj. 自顶向下;组织管理严密的
contribute to	有助于;捐献;带来,促成;
lead to	导致;通向
in addition	另外,此外
result in	导致,结果是
lack of	没有,缺乏;不足,不够

Ⅲ Exercises

1 Translate the following phrases into English

(1)交通量

(2)跑道容量

(3)战术决策

(4)关键绩效领域

(5)运行效率

(6)信息交互

(7)协同决策

(8)最小化滑行队列

(9)航空系统组块升级

(10)驻地运行

2 Translate the following sentences into Chinese

(1) Under financial and environmental constraints, there is a need to maximize assets within current budgets and with existing infrastructure.

(2) Stakeholders are tasked to attain maximum performance and service from their assets, while simultaneously delivering improved capacity and cost efficiencies.

(3) Without shared situational awareness across all stakeholders, and the ability to appropriately plan, both tactically and strategically, the delays can compound and escalate, further deteriorating the situation and leading to a blaming culture among stakeholders.

(4) If boarding of a flight will commence 15 minutes later than planned due to a

malfunctioning computer at the gate, this information will be available simultaneously to the ramp agent, gate planning, tow truck dispatcher, en-route and terminal ATCO, and possibly to entities at the affected flight's destination airport.

(5) This will allow new optimal start-up sequences to be calculated that take into account arriving aircraft, individual taxi times, de-icing requests and capacity restrictions of the surrounding airspace.

IV   Supplement Reading

Pilot: Don't fret about crowded skies

By Daniel E. Fahl

More than 5,000 flights may be in the sky at once on the busiest holiday travel days.

The aircraft operating these flights navigate intersecting airborne highways, so it's quite normal to peek out your window and witness another aircraft passing by. Passengers often ask me if pilots are aware of other nearby aircraft while in flight, and whether they ever come "too close for comfort." The answers, in order, are yes and rarely.

Pilots have many tools at their disposal that clearly paint a picture of their surroundings. Today's modern commercial aircraft are equipped with traffic avoidance systems that show the position of other aircraft. Pilots can determine from a display if nearby aircraft are above or below them, as well as if they are climbing or descending.

If the system determines that a nearby aircraft is on a collision course, the pilots will be alerted. The alert will even go as far as to direct the pilots to climb or descend their aircraft to avoid a collision. Considering the many thousands of flights each day, it is very uncommon for such a drastic step to be required.

The primary responsibility for maintaining safe aircraft separation rests on the shoulders of air traffic control, or ATC, which uses an extensive radar network that allows controllers to keep close tabs on the position of each aircraft in the sky.

Controllers keep commercial aircraft horizontally spaced by miles, and vertically spaced by a minimum of 1,000 feet. ATC has its own warning system that tracks potential conflicts of aircraft coming too close to one another. If necessary, it will provide instructions to pilots to help maneuver their aircraft out of harm's way. Such situations are normally detected well in advance, and again, happen rarely.

While air traffic controllers keep airliners safely spaced in the sky, the safety of aircraft separation begins on the ground, when flights are planned. Rather than flying randomly all over the sky, commercial flights follow published airborne highways that are defined by latitude/longitude coordinates.

With permission from ATC, aircraft may deviate from these highways to fly more directly to a destination, or to avoid weather. Altitudes are not assigned at random either. Commercial flights headed east fly at odd-numbered altitudes, such as 35,000 or 37,000 feet, while westbound flights fly at even-numbered altitudes, such as 34,000, or 36,000 feet.

When traffic congestion is high, the need to keep flights at a safe distance from one another can lead to delays. Foul weather only worsens the problem, as storms can block valuable airspace.

Delays such as these will often be communicated by the airline as an "air traffic control delay." The best way to avoid delays is to avoid peak travel hours at major airports. Opt for early morning flights. Most airlines have apps that allow passengers to keep tabs on the status of their flight while on the go.

On behalf of the thousands of professional airline pilots nationwide, rest assured that getting you home safely is our priority year-round.

Unit 7

Airport Emergency Plans

Ⅰ Reading Text

Aviation Career Profile: Airport Management and
Administration and Airport Emergency Plans

Airport managers are at the center of the airport. They are the decision-makers and policy-makers for airports. They create jobs and manage each airport department. The job is multi-faceted and vital to aviation safety. And bad management of an airport can have severe consequence to the local economy and beyond.

Airports are often one of the largest employers in a local area. Large airports like JFK can have upwards of 30,000 people employed.

Small airports may only employ an airport manager and a couple of linemen. Regardless, someone has to manage the daily operations and plan for the future, and that job lies within the "airport management" title.

An average sized airport will likely employ a few different managers, like an airport manager, a director of operations or operations manager, and department managers.

1. Airport Manager   The airport manager is often employed by the city the airport is located in, and he or she is responsible for all airport operations. The airport manager oversees all other employees and departments, and manages the day-to-day operations as well as future airport planning.

Airport managers may deal with many different issues, but are primarily responsible for airport safety, regulations and budget planning.

Managers may have to deal with noise complaint, emission testing, and airport equipment management. They work closely with the FAA and other industry groups to manage and develop air traffic procedures, install air navigation equipment, mitigate safety hazard and manage the airport budget. They must work closely with numerous people, including the FAA, NTSB, airline management, air traffic controllers, firefighters, security personnel linemen, and maintenance personnel, as well as administrative workers, food workers and sometimes retail managers.

Airport managers will generally work with city, state and federal officials to make their airport safe and efficient while following rules and regulations. They sometimes lobby for making important changes and will work with legislative officials to promote aviation.

2. Operations Manager   An operations manager works under an airport manager, but in some cases the airport manager and operations manager can be combined into a single position. The operations manager oversees the daily operation of the airports, which may include specific departments such as maintenance, line personnel, security personnel and the overall safety of the airport.

The operations manager will be familiar with all incoming and outgoing air traffic, passenger numbers and fuel usage. They will usually be responsible for implementing regulatory practices, ensuring safety manuals and procedures are up to date and that programs are planned and implemented as necessary.

Hazardous weather operations, snow removal, environmental factors (bird strike avoidance) and airport safety practices (an emergency response plan) are primary responsibilities for an operations manager.

3. Department Managers   At larger airports, there will be usually different airport departments and multiple managers. There might be a budget manager, a vehicle maintenance manager, and a food service manager. There is usually a safety program manager, an emergency response team manager and a building maintenance manager.

At really large airport, there is a multitude of management position. For example, an airport as large as Dallas/Fort Worth (DFW) is organized by a board of directors, in which each director is tasked with a different job. For example, DFW airport has directors for the following departments, each with an executive vice president in charge: Finance, Administration & Diversity, Operations, Revenue Management, Government Affairs, and Airport Development and

Engineering. Under the EVP each of these departments are vice presidents for smaller department, such as I. T. , human resources, environmental affairs, public safety, public affairs, marketing, concessions, and parking, to name a few.

In this case, all of these managers or directors will work together with the airport manager to maintain a safe, efficient airport for everyone.

Administrative Assistant   Depending on how large the airport is, there might be one or many administrative assistants. Large airports might also employ specialists, such as legal specialists, accountants, and bookkeepers.

Sometimes there are one or two assistants to the airport managers, and at other times there are administrative assistants for each department, such as maintenance, fuel, engineering, environmental, and sales departments, to name a few.

4. Airport Emergency Plans   Ever wonder how airport emergency operations work? What happens in the moment after a plane crash? Well, airports have a detailed airport emergency plan (AEP) to help everyone deal with the aftermath of an emergency or disaster.

A typical airport emergency plan involves several different components and is usually created and implemented by either the airport manager or an emergency response coordinator.

Here's a brief description of who is involved in an airport emergency response plan, and how it all works.

Parties might be involved in an AEP. There are always several parties involved in the creation and execution of an AEP. Here is a list of just a few of the people and groups that help coordinate an AEP:

- Airport Emergency Response Coordinator
- Airport manager
- Airport Rescue and Firefighting Crews (ARFF)
- Airport security team
- Air carriers and other airport occupants
- Air traffic control
- Community emergency management teams
- Local law enforcement
- Local hospitals and other medical teams
- Local or federal mutual aid organizations and relief organizations, such as the American Red Cross and FEMA

- Media outlets
- FAA
- NTSB, in the event an aircraft accident investigation is needed
- FBI, in the event of an act of terrorism or national security
- Military agencies, if available

5. Formation of an AEP   The creation of an AEP isn't a simple task.

First, research must be done to formulate the best plan based on many other plans, such as the city's emergency response plan, local ordinances, OSHA and EPA plans, regional and federal emergency response plans and even individual air carrier plans.

Second, an AEP must comply with several regulations from different agencies such as OSHA, the FAA and the Department of Transportation (DOT).

Then an analysis must be done to identify hazard of the particular airport involved with the AEP. For example, one airport might be subject to volcanic activity or tornados, while another might be in a high-risk zone for a terrorist attack.

Once potential hazards are identified and a risk assessment is completed, an airport emergency response coordinator can begin to develop plan for specific scenario. There will be a different plan for a plane crash, for instance, than for a bomb threat.

Drafting an AEP takes multiple meetings with many different groups of people, and multiple revisions before complete. Once complete, AEP testing can begin.

An AEP is always being revised. One of the things that help managers and coordinators to develop the best possible plan is to practice the plan over and over again, exhausting different scenario and utilizing all available resource to make sure all parties know their role if an emergency occurs. There are a few different methods used to test the potential success of an AEP.

1. Training   Training must be in-depth and frequent. There are many people that need to be familiar with the AEP, so generalized training manuals and classroom sessions are popular choices for training many people at once. There should also be specialized training for certain groups, depending on the each role. First responders, firefighters, airport security, and others will need specific training on how to manage injuries, crowd, and the media, as well as how to handle sensitive information while protecting the scene of the disaster.

2. Drills   Fires, bomb threats, and hazardous material handling can all be practiced with frequent drills. Drills typically focus on a single aspect of the AEP, such as how to notify everyone, how to secure the communication process, or how to handle evidence.

3. Exercises An exercise can be either a tabletop exercise, a functional exercise or a full-scale exercise.

Tabletop exercise is the most simple, as it merely involves a meeting atmosphere and a discussion of AEP limitations and improvements that could be made.

A functional exercise involves in a pretend scenario with time constraints and goals for completion but doesn't include every aspect of an AEP.

A live exercise also called a full-scale exercise, includes a live simulation of an emergency event, such as a plane crash. Full-scale exercises involve in many groups, such as emergency response teams, the Red Cross, local hotels, firefighters, police, airline operations staff, NTSB investigators, etc.

The scope of a live exercise will depend on the requirements of the airport (some airport are required to complete a full-scale exercise every three years), the type of scenario that is being rehearsed and the availability of associated groups. In many cases, it's very real, even including actors that pretend to be injured passengers, like in this full-scale exercise in Chicago.

According to an FAA advisory circular regarding guidance for AEPs, the elements of an AEP usually include the following:

- A list of the parties and the primary responsibilities of each group during and after a disaster.
- A list of key people that will be notified in the event of an emergency, and what each person's role will be.
- Notification procedures, including methods of communication and the order by which people will be notified.
- Specific checklists for different scenario.
- A description of how and when information will be disseminated to the public, including who will talk to the media and which pieces of information will be released, paying particular attention to sensitive information.
- A description of evacuation and sheltering techniques, as well as management of local and federal aid resources.
- Information on how to secure the area, letting people in and out of hazardous area and sensitive information area.
- Guidance for firefighting, health, and medical parties.
- Instructions on how and when to obtain additional resources, airport equipment

management, and safety.

- Airport map, building location, and airport ground information.

II   Words and Expressions

regardless	*adv.*	不管;不顾;不管怎样,无论如何
budget	*n.*	预算
maintenance	*n.*	维护,养护,保养;保持,维持
federal	*adj.*	联邦(制)的
legislative	*adj.*	立法的
regulatory	*adj.*	具有监管权的,监管的,管理的
revenue	*n.*	收益,收入
aftermath	*n.*	后果,余波
coordinator	*n.*	协调人,统筹者
tornado	*n.*	龙卷风
terrorist	*n.*	恐怖分子
draft	*n.*	草稿;草图;草案
revise	*vt.*	改变;修正;温习,复习
exhaust	*vt.*	使精疲力竭
scenario	*n.*	可能发生的事,可能出现的情况;剧情梗概
utilize	*vt.*	利用,使用
hazardous	*adj.*	危险的,不安全的
completion	*n.*	完成,结束
rehearse	*vt.*	(使)排练,(使)排演
hazardous	*adj.*	有危险的;冒险的;碰运气的
in charge		主管,在掌管之下,负责,看管
be subject to		服从,经受,受支配,隶属
full-scale		彻底的;全面的

be combined into　被组合成

III　Exercises

1　Translate the following phrases into English

(1) 鸟击

(2) 应急反应

(3) 机场应急规划

(4) 噪音投诉

(5) 排放测试

(6) 危险天气运行

(7) 公共安全

(8) 恐怖袭击

(9) 咨询通告

(10) 坠机

2　Translate the following sentences into Chinese

(1) The operations manager will be familiar with all incoming and outgoing air traffic, passenger numbers and fuel usage. They will usually be responsible for implementing regulatory practices, ensuring safety manuals and procedures are up to date and that programs are planned and implemented as necessary.

(2) A typical airport emergency plan involves several different components and is usually created and implemented by either the airport manager or an emergency response coordinator.

(3) Once potential hazards are identified and a risk assessment is completed, an airport emergency response coordinator can begin to develop plans for specific scenario. There will be a

different plan for a plane crash, for instance, than for a bomb threat.

(4) One of the things that help managers and coordinators to develop the best possible plan is to practice the plan over and over again, exhausting different scenario and utilizing all available resource to make sure all parties know their role if an emergency occurs.

(5) Under the EVP each of these departments are vice presidents for smaller departments, such as I. T. , human resources, environmental affairs, public safety, public affairs, marketing, concessions, and parking, to name a few.

IV　Supplement Reading

Airport and Runway Lights

If you've been to any major airport at night, you may have noticed that there are a lot of different kinds of light, ranging from flashing white or pulsating yellow to steady red and even blue. Airport lighting is obviously important for aircraft operating at night, but why do we need so many types? And what do all the colors mean? Airport lights can be divided into different types: general airport lighting, taxiway lighting, runway lighting, and approach light system.

General Airport Lighting

General airport lighting usually includes the airport beacon and any white or red beacon light on top of towers, buildings, and construction equipment. The airport beacon is a large, powerful rotating light that's highly visible from miles away. Public use airport beacon rotates green and white. Military airport rotates green and white but has two white lights for each green light, which differentiates them from civilian airport.

And heliports rotate among green, white and yellow lights. Pilots flying cross – country can easily identify an airport at night from its beacon, making it one of the easiest checkpoints for pilots when navigating at night. Sometimes air traffic controllers can turn the beacon on and off as necessary; other times it's set on a timer. Airport buildings, towers, and other tall equipment on the field will have a small, steady red beacon on top of them to aid in collision avoidance for low-flying aircraft.

Taxiway Lights

- Taxiway Edge Lights   Taxiway edge lights are blue in color and line up the taxiway. Airports often have green taxiway centerline lights as well.
- Clearance Bar Lights   Set inside the taxiway, clearance bar lights are steady yellow and are meant to increase the visibility of a hold line or a taxiway intersection at night.
- Stop Bar Lights   Only installed at selected airport, stop bar lights are meant to reinforce an ATC clearance to cross or enter into a runway in low visibility situation (low IMC). They're in-pavement lights that are steady red and extend across the taxiway at a hold short line. Once a pilot is cleared onto the runway, the stop bar lights will be turned off.
- Runway Guard Lights   A pair of two steady yellow lights are positioned at each side of the taxiway at the hold short line, and the runway guard lights are meant to draw attention to the hold short line—the area where a taxiway meets the runway.

Runway Lights

- Runway End Identifier Lights (REILs)   A pair of white flashing lights, one on each side of the approach end of the runway, that help identify the runway from taxiway at night.
- Runway Edge Light Systems (HIRL/MIRL/LIRL)   The runway edge lights are steady white light on the edges of the runway. On instrument runways, the white light change to yellow during last 2,000 feet, or half of the runway length, whichever is less, and then they turn red as the aircraft reaches the end of the runway. They can be high-intensity (HIRL), medium-intensity (MIRL) or low-intensity (LIRL).
- Runway Centerline Lighting System (RCLS)   On some precision runways, a runway centerline light system is installed, with white lights spaced at 50-ft intervals on the

centerline of the runway. With 3,000 feet remaining, the white lights change to alternating white and red, and then all red during last 1,000 feet.

- Touchdown Zone Lights (TDZL)   Touchdown zone lights are steady white lights placed in two rows next to the centerline, starting at 100 feet and extending to the midpoint of the runway, or 3,000 feet beyond the threshold, whichever is less.

- Land and Hold Short Lights   When land and hold short operations (LAHSO) are in effect, flashing white light may be seen across the runway at the hold short line.

- Other runway lighting may include Runway Status Lights (RWSL), which includes Runway Entrance Lights (RELs), the Takeoff Hold Light Array (THL), Runway Intersection Lights (RILs), and the Final Approach Runway Occupancy Signal (FAROS). These lights work in conjunction with surveillance systems (like ADS-B) and are fully automated. They assist in informing pilots and ground vehicle operators when it's safe to enter or cross a runway.

Visual Glideslope Indicators

Visual glideslope indicators are meant to give pilots a visual guide during their descent in order to maintain a stabilized approach. They come in two types, VASIs and PAPIs, each of which has multiple types of arrangements, but both of which give pilots a good idea whether they're on the glide path for a stable approach or not.

- VASIs, or Visual Approach Slope Indicators, are bars of light on the side of the runway that, when illuminated, give pilots a visual indication of whether their aircraft is too high or too low on the approach. VASIs can be made up of 2, 4, 6, 12 or 16 lights, usually located on two or three bars-near, middle and far. Two-bar VASIs provide an indication for a 3-degree glideslope, which is typical for what should be flown during an approach. In a common two-bar VASI system, a pilot should see two red lights on the far bars and two white lights on the near bars. If all lights on the near and far bars are red, he is too low. If all lights on the near and far bars are white, he is too high. The rule of thumb pilots used is "red over white, you're all right."

- PAPI stands for Precision Approach Path Indicator. PAPI lights are arranged horizontally, and typically include four lights that can be red or white, depending on where the aircraft is on the glideslope. A typical PAPI system is located on the left side of the runway. When all four lights are white, the aircraft is too high. As it descends onto the glide path, the lights on the right side will begin to turn red. When an aircraft is on the precise glide path, the two left lights should be white, and the two right lights should be red. When three or more lights are red, it indicates that the aircraft is too low.

Unit 8

Navigation System

I Reading Text

Initially, because they lacked flight instruments or navigation systems, airline pilots were limited to daylight flying during good weather condition. The pilots were forced to use outside visual references to control their aircraft's attitude, relying on the natural horizon as a reference. They would note any changes in the flight attitude of their aircraft and make the necessary control adjustments that would keep their aircraft in level flight.

Pilots navigated from airport to airport using either pilotage or deduced reckoning (commonly called dead reckoning). Pilotage required that the pilot used a map of the surrounding area as a reference. The pilot would draw a line on the map, extending from the departure to the destination airport, and note any prominent landmark that would be passed while in flight. As the aircraft passed these landmarks, the pilot would note any deviation from the planned flight path and adjust the aircraft's heading to return to the preplanned course.

Since the wind at the aircraft's cruising altitude usually caused the aircraft to drift towards either left or right of course, the pilot was forced to constantly alter the aircraft's heading to counteract these crosswind. This change in heading is known as the crosswind correction angle or wind correction angle. The resultant path in which the aircraft flies over the ground is known as the ground track or the course.

The maps used by pilots in the early 1920s were common road maps available at automobile service station. These maps were unsuitable for aerial navigation since they lacked the necessary landmark information needed to accurately navigate from one airport to the next. It soon became apparent that pilots needed a specialized chart expressly designed for using in aeronautical navigation. The U. S. government then developed and began to print such air navigation chart, known as sectional chart.

When flying using VFR rules, most pilots use dead reckoning, in combination with pilotage, to navigate to their destination. With dead reckoning, the pilot uses the forecast winds at the planned cruising altitude and applies trigonometry to deduce the proper heading that the aircraft should fly to counteract the crosswind. Properly calculated, this method of navigation is very accurate; however, it is hampered by the fact that the winds-aloft information is a forecast but not a reflection of the actual winds. To verify that dead reckoning has calculated the proper heading, the pilot must still visually check the accuracy of the deduced heading by using a sectional chart.

The first step in planning a flight using both dead reckoning and pilotage is to determine the true course that will lead the aircraft to the destination airport. This is accomplished by drawing a line from the departure airport to the destination on the sectional chart. The pilot then determines the angle of this course in reference to real north, using a device called a plotter. The pilot obtains the forecast wind speed and direction at the chosen cruising altitude and, using either a mechanical or an electronic computer, calculates the true heading that the aircraft must fly.

Aeronautical charts cannot be properly used by pilots unless they have accurate aircraft heading information. All of these charts are oriented with respect to true north. Unfortunately, the only instrument aboard most aircraft actually indicates heading is a magnetic compass, which usually points toward magnetic north.

The angular difference between true north and magnetic north is known as variation. The variation depends on the aircraft's current location. In different areas of the United States, the variation may range from 0° to as much as 20°. To properly use the magnetic compass when navigating, the pilot must add the variation or subtract it from the aircraft's true heading to determine the magnetic heading that must be flown. The pilot may then fly this heading to use the aircraft's magnetic compass.

Although the magnetic compass is a relatively reliable instrument, it is subject to various inaccuracies. One of these inaccuracies is known as deviation. Deviation is caused by the stray magnetic fields of electrical equipment or metallic structures within the aircraft. Since all aircraft contain some stray magnetic fields, every plane is required to be equipped with a compass deviation card that lists the inaccuracies and the correction that must be applied when interpreting the magnetic compass.

A few other conditions can cause the magnetic compass to indicate inaccurately. During changes in airspeed or while the aircraft is turning, the magnetic compass will not indicate correctly. These particular inaccuracies are known as acceleration and turning errors. In general, the only time that the magnetic compass can be accurately interpreted is when the aircraft is in straight and level, unaccelerated flight. In addition, the placement of a metal or magnetized object (a flashlight, clipboard, or screwdriver) near the compass will alter the local magnetic field and cause magnetic compass errors.

While the CAA was developing and installing A-N radio range, the nondirectional radio beacon (NDB) was also being developed. The NDB transmits a uniform signal Omni directionally from the transmitter, using the low-and medium-frequency band (190 ~ 540 kHz). The receiver on the aircraft (known as a direction finder or DF) was originally equipped with a loop type antenna that the pilot rotated manually. When the antenna rotated so that the plane of the loop was perpendicular to the transmitted signal, the "null" position was reached, and the pilot would be able to hear the transmitted signal. Using the magnetic compass and the NDB receiver, the pilot could then determine the aircraft's bearing from the nondirectional beacon. This bearing could be plotted on a chart as line of position. Plotting lines of position from two NDBs permitted the pilot to pinpoint the aircraft's exact location. If the pilot wished to fly toward the NDB, he or she would turn until the NDB station was located directly ahead of the aircraft. If the winds aloft caused the aircraft to drift off course, the pilot would readjust the aircraft's heading, keeping the NDB directly ahead of the aircraft. This method of navigation is called homing.

Trying to manually manipulate the DF antenna while flying the aircraft is proved to be a cumbersome method of navigation and usually provided the pilot with relatively inaccurate position information. As advances were made in aircraft electronics, the manually operated NDB receiver was soon replaced by the automatic direction finder (ADF), which could electronically determine the bearing to the NDB and display this information to the pilot. Using ADF equipment in conjunction with the aircraft's heading indicator, the pilot could easily determine the aircraft's relative bearing from the station and use this information to determine the proper heading that would lead to the beacon.

The Washington Institute of Technology delivered the first operable VHF omnidirectional range (VOR) to the CAA in 1944. Each VOR is assigned to a frequency between 108.10 and 117.90 MHz. The VOR transmission is modulated with two signals: a reference-phase signal that is constant in all directions and a variable-phase signal whose phase varies with azimuth. The variable – phase signal is modulated so that at magnetic north the reference and variable signals are precisely in phase with each other. In any other direction, the VOR is designed in order that the two signals are no longer in phase.

The VOR receiver on the aircraft measures the phase difference between the two signals to determine the azimuth angle of the aircraft in relation to the VOR transmitter. The radial to be flown by the pilot is selected on the aircraft's VOR indicator using the omni bearing selector (OBS). After selecting the appropriate VOR frequency, the indicator in the cockpit will inform the pilot whether the selected course will lead to the station or away from it (known as the To-From flag). The VOR indicator will also display any lateral deviation from the selected course, using a vertical pointer known as the course deviation indicator (CDI).

The VOR provides only bearing information to the pilot (known as rho), not distance from the station (known as theta). There are only two ways for a pilot using the VOR to accurately determine an aircraft's position: using either rho-rho or rho-theta position determination. Rho-rho position determination requires that the pilot obtain bearing information from two different VORs. Using airborne VOR equipment, the pilot can plot a line of position from each VOR. These two lines of position (or radials) are then plotted on a navigation chart, with the aircraft being located at the intersection of the two radials.

If a pilot wishes to determine an aircraft's location using just one station, rho-theta position determination techniques must be used. The pilot must determine on which radial the aircraft is located (rho) and then use distance measuring equipment (DME) to determine the aircraft's distance (theta) from the VOR transmitter. Rho-theta position determination requires specialized DME equipment both on the aircraft and at the VOR transmitter.

The DME system uses the principle of elapsed time measurement as the basis for distance measurement. The DME system consists of an interrogator located on the aircraft and a transponder located at the ground station. At regularly spaced intervals, the interrogator transmits a coded pulse on a frequency of around 1,000 mHz.

When the ground-based DME transponder receives this pulse, it triggers a coded reply that is transmitted on a different frequency. When the interrogator receives this pulse, the elapsed range time is electronically calculated. Range time is the interval of time between the transmission of an interrogation and the receipt of the reply to that interrogation. The approximate range time for a signal to travel 1 nautical mile and return is 12.36 microseconds. The DME equipment on board the aircraft measures the elapsed time between interrogator transmission and reception of that signal. This time is divided by 12.36 microseconds, providing the distance that the aircraft is from the ground station.

This determination is known as the line of sight or slant range distance. Slant range is the actual distance between the aircraft and the ground based DME transponder. As the aircraft's altitude increases, the difference between slant range and ground distance increases. For instance, if an aircraft is 5.0 ground miles from the DME station, at an altitude of 6,000 feet, the DME indicator on the aircraft will indicate approximately 5.1 nautical miles from the station. But if the aircraft is directly over the DME station, at an altitude of 30,000 feet, the DME indicator will also indicate about 5.1 nautical miles.

The Global Navigation Satellite System (GNSS) is the accepted term for navigation systems that provide ground-based users with global navigation via space-based satellite systems. GNSS transmitters are typically located on Low Earth Orbit satellites permitting users with fairly small, inexpensive receivers to determine their location in three dimensions (latitude, longitude and altitude). As long as the transmitters are within the sight line of a number of satellites, the receivers can determine their location within a few meters or even feet.

PBN is not a navigation system but a framework for defining a navigation performance specification within which aircraft must comply with specified operational performance requirements. Unlike other navigation specifications, PBN is not specific equipment but rather establishes required performance on the basis of defined operational need. It is the aircraft's own capability that determines whether the pilot can achieve the specified performance and qualify for the specific operation.

II Words and Expressions

adjustment   *n.* 调整,调节;调节器

pilotage	*n.* 领航;领航费;驾驶术
extend	*n.* 扩展;伸长
	vt. 延伸;扩展
prominent	*adj.* 突出的,显著的;杰出的;卓越的
landmark	*n.* 陆标;地标;界标;里程碑
deviation	*n.* 偏差;误差
drift	*n.* 漂流,漂移;趋势;漂流物
counteract	*vt.* 抵消;中和;阻碍
	vi. 漂流,漂移;漂泊
	vt. 使……漂流;使……受风吹积
course	*n.* 科目;课程;过程;进程;路线,航向
	vt. 追赶;跑过
	vi. 指引航线;快跑
aerial	*adj.* 空中的,航空的;空气的
	n. 天线
cruising	*adj.* 巡航的
	n. 巡航
trigonometry	*n.* 三角法
hamper	*v.* 阻碍;限制
	adj. 受阻碍的
angular	*adj.* 角度上的
metallic	*adj.* 金属的,含金属的
clipboard	*n.* 剪贴板;附有纸夹的笔记板
screwdriver	*n.* 螺丝刀
omnidirectionally	*adv.* 全方向地
antenna	*n.* 天线
perpendicular	*adj.* 垂直的,正交的;直立的;陡峭的
	n. 垂线;垂直的位置
pinpoint	*vt.* 查明;精确地找到;准确描述
	adj. 精确的;详尽的
homing	*adj.* 自动导引的;有返回性的

	n. 归航;动物的返回性
	vt. 回家
manipulate	*vt.* 操纵;操作;巧妙地处理;篡改
cumbersome	*adj.* 笨重的;累赘的;难处理的
beacon	*n.* 灯塔,信号浮标;烽火;指路明灯
	vt. 照亮,指引
	vi. 像灯塔般照耀
azimuth	*n.* 方位;方位角;偏振角
modulate	*vt.* 调节;调制;调整
intersection	*n.* 交叉;十字路口;交集;交叉点
radial	*adj.* 半径的;放射状的;光线的;光线状的
	n. 射线,光线
interrogator	*n.* 质问者;询问机
transponder	*n.* 异频雷达收发机;转调器,变换器
pulse	*n.* 脉冲;脉搏
	vt. 使跳动
	vi. 跳动,脉跳
elapsed	*v.* 时间过去;消逝(elapse 的过去分词)
	adj. 过去的;经过的
specification	*n.* 规格;说明书;详述
sectional chart	截面图
in combination with	与……结合,与……联合
nondirectional radio beacon	无向信标台
direction finder	探向器;方位仪
loop type	循环类型
range time	时间范围
low earth orbit	近地轨道
on the basis of	根据;基于…

III Exercises

1 Translate the following phrases into Chinese

(1)推测领航

(2)侧风修正角

(3)近地轨道

(4)地面航迹

(5)真航向

(6)磁北

(7)磁航向

(8)磁罗盘

(9)自动定向仪

(10)指南针偏差

2 Translate the following sentences into Chinese

(1) As the aircraft passed these landmarks, the pilot would note any deviation from the planned flight path and adjust the aircraft's heading to return to the preplanned course.

(2) During changes in airspeed or while the aircraft is turning, the magnetic compass will not indicate correctly. These particular inaccuracies are known as acceleration and turning errors.

(3) In addition, the placement of a metal or magnetized object (a flashlight, clipboard, or screwdriver) near the compass will alter the local magnetic field and cause magnetic compass errors.

(4) As advances were made in aircraft electronics, the manually operated NDB receiver was soon replaced by the automatic direction finder (ADF), which could electronically determine the bearing to the NDB and display this information to the pilot.

(5) When the ground-based DME transponder receives this pulse, it triggers a coded reply that is transmitted on a different frequency. When the interrogator receives this pulse, the elapsed range time is electronically calculated.

IV   Supplement Reading

Airline Navigation Threatened By Plasma Plumes

By David Shiga

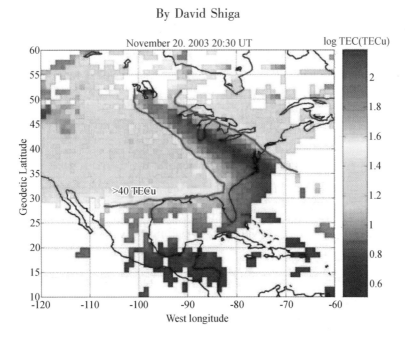

Figure 1   A plasma plume over North America was mapped on 20 November 2003 by its effect on GPS signal

Mysterious plumes of plasma at the edge of the Earth's atmosphere are threatening airline navigation by throwing off GPS positioning information by up to the length of a football field, but a poor understanding of how the plumes form means that accurate forecasts of the phenomena— which would allow GPS users to plan around them—are years away.

Scientists have long understood that outbursts from the Sun called coronal mass ejection can interfere with communication between Earth-orbiting satellites and the ground.

More recently, they have learned that at least some of the disruptions are due to giant plumes of charged particles, or plasma, that form in response to the solar outbursts in the Earth's ionosphere, a region filled with ions at the boundary between Earth's atmosphere and outer space. The concentrated plasma in the plumes diverts and delays satellite communication, such as GPS signal[Figure 1].

However, forecasting these disruptions is difficult because scientists do not know exactly how the plumes form or where the extra plasma they contain comes from.

Real-time maps

There are hints that the ionosphere above the equator could be the source of the plume plasma, feeding it outward to other parts of the ionosphere where the plumes are generated. However, it has been difficult to track ionosphere activity at the equator because of a lack of sensors there.

Now scientists are trying to organise the creation of a network of GPS receivers around the equator in Africa that will calculate the distortion that the ionosphere causes to signals sent by GPS satellites. In this way they hope study the plumes and hunt for their source.

The researchers are discussing on the project this week at the Africa Space Weather Workshop in Addis Ababa, Ethiopia.

"North America has thousands of GPS receivers in a network we use to monitor North American plumes," says Tim Fuller-Rowell of the University of Colorado in Boulder, US, who helped organise the meeting in Ethiopia. The data from the receivers is used to produce online maps of the ionosphere and update every 15 minutes.

Africa has only a few dozen of dedicated receivers, Fuller-Rowell says, but that may soon change. "Within a few years we hope to deploy hundreds of receivers," he says. "Five years from now we hope to be making real-time maps of the ionosphere over Africa, too."

Satellite imaging

In addition to disrupting satellite transmission, the plume can disrupt airline navigation and radio communication, Fuller Rowell says.

Anthony Manucci of NASA's Jet Propulsion Laboratory in Pasadena, California, US, says the measurement of the ionosphere above the equator could be important in understanding how the plumes are created.

In addition to the African GPS network, scientists would like to send up a satellite to monitor the plumes from above. An ultraviolet camera on such a satellite could provide better image of the plumes than those produced from GPS measurement, Manucci says.

"If we could go up to space and put an imager looking down, we could see these plumes form and have a much better picture of why they formed in the first place," he says.

Unit 9

Automated Dependent Surveillance-Broadcast

I   Reading Text

How ADS-B Works: A Look at the Foundation of NextGen

Overview

ADS-B stands for Automatic Dependent Surveillance-Broadcast. ADS-B is the foundation of the FAA's Next Generation Transportation System (NextGen). It was developed to help transform the nation's airspace system into more efficient one. The air traffic system will undergo a much-needed modernization plan through the implementation of NextGen, and ADS-B is the primary component.

The main role of ADS-B is to provide precise aircraft location information to air traffic controllers. It's a step above radar, which has been in use for years. It uses GPS satellite signal to broadcast aircraft information continuously to air traffic controllers and other participating aircraft. ADS-B is the most accurate surveillance system that the aviation industry has ever seen. It will enable aircraft to fly more direct routes, ease congestion, decrease carbon emission and save aircraft operators' time and money.

Components

- GNSS Satellite Constellation: ADS-B is a satellite-based system. Data is continuously sent from the set of satellites to the aircraft's onboard GPS devices, where it's interpreted and then sent to ADS-B ground stations.
- Ground Stations: there will be at least 700 ground stations in the United States that receive satellite data and transmit the data to air traffic control stations.
- IFR Certified, WAAS (Wide Area Augmentation System)-enabled GPS receiver: aircraft must be equipped with a compatible GPS receiver for ADS-B to work.
- A 1,090 MHz Extended Squitter link with a Mode S transponder or a 978 MHz Universal Access Transceiver (UAS) for use with an existing transponder: the latter option is available for aircraft flying below 18,000 feet in the United States.

How it Works

ADS-B works by using satellites signals and aircraft avionics systems to interpret aircraft data and broadcast it to air traffic controllers on a continuous basis and in almost real-time.

Satellite signals are interpreted by an aircraft GPS receiver. ADS-B technology takes the satellite data and additional data from aircraft avionics to create a very accurate picture of the aircraft's location, speed, altitude, and over 40 other parameters. This data is transmitted to a ground station and then to air traffic controllers. Other properly equipped aircraft in the area will also receive the data, increasing situational awareness for pilots.

There are two different functions of ADS-B: ADS-B In and ADS-B Out.

- ADS-B Out is the first and main function that the FAA has addressed. An aircraft that is capable of ADS-B Out has the capability of broadcasting its position, speed and altitude to air traffic controllers and other ADS-B equipped airplanes. According to an FAA mandate, all aircraft that wants to fly in airspace and currently requires a transponder must be equipped with ADS-B Out capabilities before January 1, 2020.
- ADS-B In remains an optional capability at least for now. The ADS-B In capability will allow aircraft to receive traffic and weather information in real-time on the aircraft cockpit display. The ADS-B In function goes above and beyond today's traffic systems

(such as TCAS) as it offers more precise data and more data parameters than current TCAS systems do. For example, TCAS can display the vertical distance from aircraft but not lateral. ADS-B In will display the speed, locations, altitude and vectors of other participating aircraft, along with many other pieces of data.

Errors and Limitations

Currently, the biggest limitation for ADS-B is the cost of installing the necessary equipment on virtually every aircraft in the country. While the program makes flying safer and more efficient, most flight departments and general aviation pilots are having a difficult time justifying the cost.

ADS-B has very few system errors; in contrast, it is known for its reliability. No human-made system is fool-proof, though, and some experts claim that ADS-B (and GPS in general) is vulnerable to system infrastructure attacks such as hackers or GPS jamming. Additionally, since ADS-B is reliant on the GNSS system, normal satellite errors such as timing errors and satellite weather errors can affect ADS-B.

Safety Risks

Overall, the ADS-B system is a major improvement for the future of our airspace system, but it's not without risk. With the current radar system being a mostly risk-free, accurate navigational system, a move to a completely new system brings up questions of reliability, safety risk and cost. What are those questions and risks, and have they been mitigated to an acceptable level?

While the FAA has demonstrated that the final result will be an unequivocally safer and more efficient air travel system, they've conducted research to back up their stance and they'll need to continue to examine and re-examine the program from a safety perspective. The implementation of any new system is likely to bring unknown errors and hazards.

For ADS-B, these hazards include:

- Training and human factors
- GPS failure
- Avionics malfunctions
- Security issues.

These issues have not yet to be resolved entirely, but they've been identified as risks and the measures have been taken to minimize their risk as much as possible. A 2,000 study completed a typical system of safety precedence sequence with regards to the system as a whole, and found the residual risk to be "controlled to an acceptable level."

Early in the development of ADS-B, the Capstone System Safety Working Group was established in partnership with the FAA to provide the necessary research and a preliminary hazard analysis of ADS-B. The hazards determined include the following.

Human Factors

- Confusion and loss of situational awareness
- Inappropriate use of avionics
- Pilot procedural errors
- Coordination issues with ATC
- Loss of situational awareness due to too much "heads-down" time

Ground System Risks

- Calibration errors
- Loss of communication
- Malfunction

Avionics Failure

GPS Errors

Weather, traffic and terrain malfunctions

- Lack of coverage
- Limited forecast
- Limited reporting station

Security vulnerabilities

- Spoofing, jamming and masking

For the most part, these risks have been researched, analyzed, mitigated and accepted. However, one of the biggest hazard associated with ADS-B still remains: human error. If the pilot doesn't fully understand the equipment he or she is using, the system becomes a hazard instead of a benefit. Studies suggest that advanced avionics systems require in-depth training and understanding for operators to use safely, and many operators will not voluntarily receive the training they need to safely fly with ADS-B. And the FAA's ADS-B mandate for all aircraft to be equipped with ADS-B by 2020 will intensify the cost and hazard associated with advanced avionics and human error.

The Capstone project determined that excessive heads-down time while using ADS-B would potentially result in a frequent loss of situational awareness, and although an accident might be rare in this case, a resulting accident would likely be catastrophic. This is a constant risk that will continue to be a problem for ADS-B users as it becomes a familiar addition to the flying world. Pilots must accept responsibility for mitigating this risk as much as possible through the training and awareness.

When all is said and done, ADS-B is a safe, efficient addition to the nation's airspace system. But like any navigation aid or avionics system, it's only as safe as its operator.

Current Status

According to the FAA, the organization completed all network ADS-B sensors.

These stations offer weather service and traffic information to ADS-B equipped aircraft across 28 TRACON facilities. Of 230 ATC facilities, over 100 are currently using ADS-B, with the rest expected to be fully equipped by 2019. The FAA stands by its mandate that all aircraft operating in designated airspace must be ADS-B Out equipped by January 1st, 2020.

Practical Use

There is uncertainty centered on the specific types of equipment needed for different aircraft and operators. The equipment installation varies to depend on the type of flying and currently installed equipment.

A 978 MHz UAS link, for example, will suffice for an aircraft with a WAAS-enabled, IFR certified GPS unit and a Mode C transponder have already installed, unless the operator would like to fly outside of the United States or above 18,000 feet, in which a 1,090 MHz ES link would be necessary. But a 1,090 MHz ES link isn't compatible with TIS-B or FIS-B, this means an operator would have to find another way to get traffic information (TCAS).

And an operator who doesn't have a WAAS-enabled GPS unit in his aircraft will have to purchase a new GPS unit along with a 978 MHz UAS or 1,090 MHz ES link and potentially a Mode C or Mode S transponder.

Once in use, ADS-B is a valuable tool, providing the most accurate data to air traffic controllers and pilots that we've ever seen. When it is implemented nationwide, the benefits are positive.

There's no arguing that ADS-B is quite expensive and complicated, though. The FAA is hopeful that the long-term benefits will outweigh the cost, but the project leaves aircraft owners in a difficult position.

Ⅱ Words and Expressions

undergo	*vt.*	经历,经受,遭受
surveillance	*n.*	监视
interpret	*vt.*	解释;口译,作口译
parameter	*n.*	参数; 界限
compatible	*adj.*	相容的,兼容的,可共存的,可兼用的
squitter	*n.*	断续振荡器
transponder	*n.*	应答器;应答机;转发器
transceiver	*n.*	无线电收发两用机
avionics	*n.*	航空电子设备;航空电子学

capable	*adj.*	有能力做某事,能力强的,能干的
mandate	*vt.*	授权;授权令,命令,训令;委任统治权,托管权
vertical	*adj.*	垂直的;直立的,纵向的,纵向结构的
lateral	*adj.*	侧面的;横(向)的
virtually	*adv.*	事实上
mitigate	*vt.*	减轻,缓解,缓和
intensify	*vt.*	加强;强化
unequivocally	*adv.*	明确地,毫不含糊地
stance	*n.*	观点,立场,看法
implement	*vt.*	执行;贯彻;实施
hazard	*n.*	危险,大胆猜测
malfunction	*n.*	故障,失灵;功能障碍
precedence	*n.*	优先;优先权;优先地位
sequence	*n.*	顺序,次序;一连串相关事件;一段情节,片段,连续镜头
residual	*adj.*	残余的,剩余的
calibration	*n.*	校准;刻度;准确标示
jam	*vt.*	胡乱地塞,干扰
capstone	*n.*	(墙、建筑物等的)顶石
catastrophic	*adj.*	灾难性的,失败的
suffice	*vi.*	足够,无须多说
as a matter of fact		实际上;事实上
in contrast to		与……对比或对照;与……相反
loss of control		失控
prepare ourselves for		使我们为……做好准备
at first blush		乍一看;乍一想;经初步考虑;据初步印象
associated with		与……有关系
navigation aid		航标
centered on		集中在……;以……为中心
in contrast		相反
back up		证实,备份,倒车

Ⅲ　Exercises

1　Translate the following phrases into English

(1) 目视飞行规则

(2) 仪器飞行气象条件

(3) 碳排放

(4) 广域增强系统

(5) 能见度不良

(6) 态势感知

(7) 交通信息广播服务

(8) 飞行航线

(9) 控制在可接受的水平

(10) 电子设备失效

2　Translate the following sentences into Chinese

(1) It will enable aircraft to fly more direct routes, ease congestion, decrease carbon emission and save aircraft operators time and money.

(2) ADS-B works by using satellites signals and aircraft avionics systems to interpret aircraft data and broadcast it to air traffic controllers on a continuous basis and in almost real-time.

(3) The Capstone project determined that excessive heads-down time while using ADS-B would potentially result in a frequent loss of situational awareness, and although an accident might be rare in this case, a resulting accident would likely be catastrophic.

(4) Pilots must accept responsibility for mitigating this risk as much as possible through training and awareness.

(5) Once in use, ADS-B is a valuable tool, providing the most accurate data to air traffic controllers and pilots that we've ever seen. When it is implemented nationwide, the benefits are

positive.

IV   Supplement Reading

The Differences Between ADS-B Out
and ADS-B In and How Does FIS-B Work

Automated dependent surveillance-broadcast equipment (ADS-B) allows air traffic controller and other participating aircraft to receive extremely accurate information about an aircraft's location and flight path, which in turn, allows for safer operation, reduced separation standard between aircraft, more direct flight routes, and cost savings for operators.

The ADS-B system is the foundation of the FAA's Next General Air Transportation System (NextGen).

It's a satellite-based system that was implemented as an improvement over radar to make the nation's airspace more efficient.

Two types of ADS-B can be installed on an airplane: ADS-B Out and ADS-B In. Both are valuable, but only ADS-B Out is mandated by the FAA which was established in 2010, and all aircraft operating in designated airspace must be equipped with ADS-B Out by January 1, 2020.

Note that only ADS-B Out is required in airplane, but it's important to understand how ADS-B Out and ADS-B In work together.

ADS-B Out

ADS-B Out is the broadcast part of ADS-B. An aircraft equipped with ADS-B Out capability will continuously transmit aircraft data such as airspeed, altitude, and location to ADS-B ground station. The minimum equipment needed for ADS-B Out capability includes an ADS-B approved transmitter—either a 1,090 MHz Mode S transponder or a dedicated 978 MHz UAT for use with a previously installed Mode C or Mode S transponder—and a WAAS-enabled GPS system.

ADS – B In

ADS-B In is the receiver part of the system. ADS-B In equipment allows aircraft, when equipped properly, to receive and interpret other participating aircraft's ADS-B Out data on a computer screen or an Electronic Flight Bag in the cockpit. The ADS-B In function requires an approved ADS-B Out system, along with a dedicated ADS-B receiver that has "in" capability.

Additionally, an ADS-B compatible display interface will be needed for graphic weather and traffic displays (called TIS-B and FIS-B). Other Helpful Information:

- TIS-B is short for Traffic Information Service-Broadcast. TIS-B services will work with both the 1,090 MHz Mode S transponder and the 978 MHz UAT system. No subscription service or added fees are associated with TIS-B.

- FIS-B, short for Flight Information Services-Broadcast, can be received only through the 978 MHz UAT structure. It's also free to anyone with a 978 UAT.

- A 1,090 MHz Mode S transponder is required for aircraft flying at 18,000 feet or higher and is the standard in Europe.

- A 978 MHz UAT is primarily marketed to general aviation pilots, as it can only be used below 18,000 feet in the United States.

- It is expected that aircraft owners will spend at least MYM5,000 to equip their aircraft with ADS-B Out, but costs are likely to be much higher than that.

- All aircraft that will fly into designated and controlled airspace after Jan. 1, 2020 will be required to have at least ADS-B Out capability. At this time, ADS-B In is still an optional but helpful tool for situational awareness.

- Many industry experts have called for an extension to the deadline, as well as exceptions to the rule, to have time to equip their aircraft. Maintenance facilities expect a backlog of ADS-B-related work, and many experts say there's very little chance that all of the required aircraft will be able to become ADS-B equipped before the 2020 deadline.

FIS-B is a data broadcasting service that works along with ADS-B to allow aircraft operators to receive aeronautical information such as weather and airspace restrictions through a data link to the cockpit. Along with its partner system TIS-B, FIS-B is available at no cost to ADS-B users as part of the FAA's Next Generation Air Transportation System.

The system gathers information through the use of ADS-B ground station or radar and delivers that data to an aircraft's onboard cockpit displayed in the form of weather alert, airport information and various other report. FIS-B was created for use by general aviation pilots.

How it Works

Information for FIS-B is transmitted from ground station to participated ADS-B In aircraft on a 978 MHz UAT data link. Aircraft that use a 1,090 MHx Extended Squitter transponder will not be eligible to receive the FIS-B product.

There are currently over 500 operation ground stations that are part of the ADS-B network, and the FAA is working to add approximately 200 additional stations.

The aircraft's ADS-B receiver (known as ADS-B In) interprets the data and displays it onto a screen in the cockpit. The actual interface on which FIS-B is displayed can vary, but it will typically be incorporated into a flight management system or an electronic flight bag.

Equipment

Aircraft that want to receive FIS-B information must be equipped with ADS-B Out and ADS-B In equipment. ADS-B requires a WAAS-enabled GPS receiver and a transponder when one is not already included with the ADS-B unit.

While Traffic Information Service-Broadcast(TIS-B) is available to both 978 MHx UAT and 1,090ES transponder users, FIS-B is only broadcast to ADS-B users with a dedicated 978 MHz Universal Access Transceiver (UAT).

FIS-B is not available to aircraft operators that use a 1,090ES transponder for ADS-B. Operators using a 1,090ES transponder will have to get their weather services and graphics from a third-party source, such as XM WX Satellite Weather.

A compatible cockpit display (CDIT) is also needed to display the FIS-B data in an usable format.

Limitations

FIS-B is strictly an advisory service and is not meant to take the place of standard weather briefings and preflight planning. It is not a substitute for official weather sources such as air traffic control, flight service stations, NOAA or DUATS.

FIS-B data link services operate in line-of-sight only. Aircraft receivers must be within the service volume of the ground station to receive FIS-B.

One of the advantages for pilots using the 978 MHz UAT is that basic FIS-B services will be available to use at no cost, and these services are comparable to an XM weather subscription service.

Currently, FIS-B offers the following communication services:

- Aviation weather products such as METARs, TAFs, winds aloft and NEXRAD precipitation maps.
- Temporary flight restrictions (TFRs) and status updates for special use airspace (SUA).
- AIRMETs, SIGMETs and convective SIGMETs.
- Pilot Reports (PIREPs).
- NOTAMs (Distant and FDC).

Future services might include cloud top report, lighting and turbulence information, and icing forecast in both textual and graphical depictions. It's expected that the upgraded service will originate from a third-party and might require a subscription fee.

All of the services above are updated as they become available and are transmitted every five or ten minutes, depending on the type of information. NEXRAD will be rebroadcast every 2.5 minutes.

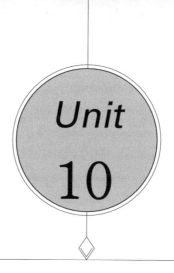

Unit 10

National Airspace System

I   Reading Text

The National Airspace System Explained and Types of Altitudes in Aviation

Airspace, Air Traffic Control and the Technology That Makes It Work

The national airspace system (NAS) was created at the dawn of commercial aviation to get aircraft from point A to point B in a safe and efficient manner. It's an old system, but it has worked for us since World War II. In fact, the United States has the safest sky in the world with respect to air transportation.

There are around 7,000 aircraft in the sky above America at once, according to the Federal Aviation Administration (FAA). This number is expected to only increase over the next 15 years, and it continues to get more difficult to fit all these aircraft into our current airspace structure. The FAA's Next Generation Air Transportation System (NextGen) promises to transform the current airspace system to optimize the use of airspace, reduce emissions, save fuel and decrease flight delay. Until NextGen is fully implemented, our current airspace system will have to suffice, though.

Airspace

The FAA classifies airspace in one of four categories:

- Controlled airspace-the airspace around busy airports, along aircraft routes, and above 18,000 feet. The FAA further divides this airspace in classes A, B, C, D and E airspace, each having different dimensions and rules.
- Uncontrolled airspace-any airspace that isn't controlled.
- Special-use airspace-restricted, prohibited, warning and alert areas, as well as military operations areas (MOAs).
- Other airspace-airspace used for temporary flight restriction.

Air Traffic Control Centers

The NAS involves more than just the control tower at your local airport. On a typical flight, a pilot will communicate with controllers at each of the following places:

- ARTCC—The airspace over the United States is divided into 22 regional sectors, each controlled by an Air Route Traffic Control Center, or ARTCC. As a flight crosses the boundary from one ARTCC region to another, the air traffic controller transfers the communication responsibility for that flight to the ARTCC controller in the next region.
- TRACON—Terminal Radar Approach Control (TRACON) is known simply as "approach" to pilots. When an aircraft gets close to an airport, the ARTCC controllers will transfer the communications to a TRACON controller, who will assist the aircraft for the arrival portion of its flight.
- ATCT—Controllers in the local air traffic control tower (ATCT) are responsible for aircraft in the associated airport's traffic pattern. Once the aircraft enters the local airport traffic pattern area, it is handed off to the ATCT, where the controllers will oversee its final approach and landing. Ground controllers are also a part of the ATCT, supervising taxi and gate operations.
- FSS—There are currently six flight service stations (FSS) in operation. Flight service specialists assist pilots with preflight planning, weather briefings, and other information pertinent to a pilot's route of flight.

Technology

In addition to the many different technologies that have been in use for years, the aviation industry is continuously developing new technologies to make the system more efficient, easier and safer for pilots and controllers. Here are just a few of them:

- Radar—Currently, the NAS relies heavily on ground-based radar system to run smoothly. The ground radar emits radio waves, which reflect off aircraft. The signal from the aircraft is then interpreted and sent digitally to computer screen at the ARTCC, TRACON or ATCT.
- Standard radio—Pilots and controllers communicate directly with VHF (very-high frequency) and UHF (ultra-high frequency) radio.
- CPDLC—Controller Pilot Data Link Communication, as the name implies, is a method for controllers and pilots to communicate via a data link. This type of communication is convenient where radios are not available and also decrease radio congestion.
- GPS—A type of navigational aid, the Global Positioning System is aviation's most accurate and most popular means of air navigation and the bread and butter of the NextGen program.
- ADS—B – In recent years, a system called ADS-B (Automatic Dependent Surveillance-Broadcast) has become popular as a mean to assist pilots and controllers in gaining a more accurate picture of air traffic, weather, and terrain during a flight.

The Next Generation Air Transportation System

Our current air traffic system gets airplanes where they need to go in a safe and organized manner, utilizing technology both old and new. While our current national airspace system has worked well for many years, it is hardly optimal for the volume of air traffic in our skies today. We are seeing more crowded runway, airport delay, wasted fuel and lost revenue than ever before. There's hope, though: The NextGen program is meant to improve the current NAS by finding methods to deal with the increased traffic and improve the overall system.

When it comes to flying airplanes, there are numerous types of altitudes where you, as a pilot, must be aware of ensuring flight safety. If you're new to flying, think of piloting as being equivalent to what baking is to cooking. A chef can play around with different recipes for his or her Bolognese sauce, but a pastry chef (much like a chemist) must follow exact instructions to bake a souffle, or else it will fall.

There are five common methods for measuring altitude in an aircraft, and each has its own applications and limitations.

True Altitude

True altitude is the height of the airplane above Mean Sea Level (MSL), a value that represents the average sea level (because actual sea level is variable). True altitude is similar to what you might call elevation in non-aviation contexts.

Most personal aircraft is not equipped to measure true altitude, so it is not used to indicate a plane's altitude. However, area forecasts (FAs) report cloud height in MSL, or true altitude. Also airport elevations, terrain, and obstacle clearance altitudes listed on the visual flight rules (VFR) sectional charts are often given in MSL.

Indicated Altitude

Indicated altitude is what is indicated on the altimeter in your airplane. It is an approximation of true altitude as measured by the altimeter. The altimeter is a basic flight instrument that measures the atmospheric pressure at the airplane's flight altitude and compares it to a preset pressure value.

The preset pressure value typically is based on the nearest weather reporting site. However, because the weather site is on the ground (and doesn't move with the plane), the pressure reported at the site may differ from the pressure at the plane's actual location, affecting the accuracy of the altimeter reading.

Indicated altitude is used to gauge a plane's distance from ground obstacles and terrain as well as vertical distances to other planes in the area, known as vertical separation. Using indicated altitude to gauge vertical separation is a relatively accurate (assuming all of the planes in a given area are set to the same weather station), but this practice is used only at altitudes under 18,000 feet.

Pressure Altitude

Pressure altitude is the altitude above the standard datum plane, a theoretical level indicated by an altimeter set to 29. 92 "Hg, the standard pressure setting". Pressure altitude is measured with barometric pressure, and a plane's altimeter is essentially a fine – tuned barometer.

Pressure altitude is important when it comes to computing aircraft performance data, including such things as takeoff and landing distance. It's also the altitude that operators use while flying above 18,000 feet or in Class D airspace, which requires everyone in flight to set their altimeters to 29. 92 "Hg in order to standardize the indicated altitudes". You can actually determine the pressure of the air by calculating the difference between the pressure altitude and the current altimeter setting.

Density Altitude

Density altitude is important for determining the performance of an aircraft, or how the aircraft will behave under certain conditions. Density altitude is pressure altitude corrected for nonstandard temperature. Because the temperature constantly changes (and is therefore nonstandard), it's very important for pilots to know the density altitude.

Density altitude is not an indication of altitude above the ground or above sea level. Rather it is a measure of the air density in a given location at the present temperature. Air density decreases with altitude; there's less air to breath at 5,000 feet than at sea level. Cold air is denser than warm air. In denser air, airplane wings have more lift, and airplane engines are more powerful because there is more oxygen to burn. As air density decreases (density altitude increases), pilot must compensate their air speed, take off and landing distance, and other factors to maintain safety.

At sea level, the standard temperature for air density calculation is 15 ℃. The surface temperature decreases on average about 2 degrees per 1,000 feet increase in elevation. For example, an airport in Colorado at 5,000 feet in elevation will have a standard temperature of 5 ℃. However, if the actual temperature at that airport is above the standard temperature, the density altitude will be higher than the normal, and airplanes might perform as though they're at, for example, 7,000 feet instead of 5,000 feet.

Absolute Altitude

Absolute altitude (AGL) is the exact height above ground level, or the actual height above the earth's surface. It is measured by a radar altimeter, which uses radar signal to measure actual distance from the ground to the aircraft. METARs and TAFs report cloud cover in AGL. Absolute altitude also is used to help land large aircraft equipped with radar altimeter. Most small aircraft do not have radar altimeter which must be substituted with indicated altitude and chart for instrument (IMC) flying and other operations.

II Words and Expressions

boundary	*n.* 分界线;边界
smooth	*adj.* 平滑的,光滑的
terrain	*n.* 地形,地势;领域;地带
utiliz	*vt.* 利用
optimal	*adj.* 最佳的;最适宜的
equivalent	*n.* 等价物,相等物
	adj. 等价的,相等的;同意义的
pastry	*n.* 油酥面团,油酥点心
souffle	*n.* 吹风声或低沉、连续不清的声音
elevation	*n.* 海拔,提升,增加,升高
altimeter	*n.* 高度计
gauge	*n.* 测量仪器,计量器,量规
barometer	*n.* 气压计,晴雨表;指标,标志
indication	*n.* 迹象;标示
decrease	*vt.* 使变小,使减少
compensate	*vt.* 补偿,赔偿;付报酬

	vi. 补偿,赔偿;抵消
substitute	*n.* 代用品;代替者
	vi. 替代
	vt. 代替
handed off	不可触摸;切换,转换
vertical separation	垂直隔离
barometric pressure	气压,大气压
density altitude	密度高度
absolute altitude	绝对高度

Ⅲ   Exercises

1   Translate the following phrases into English

(1)管制空域

(2)特殊用途空域

(3)联邦航空局

(4)飞行服务站

(5)航路管制中心

(6)终端区雷达进场管制

(7)平均海平面

(8)广播式自动相关监视

(9)全球定位系统

(10)超高频

2   Translate the following sentences into Chinese

(1) In addition to the many different technologies that have been in use for years, the

aviation industry is continuously developing new technologies to make the system more efficient, easier and safer for pilots and controllers.

(2) The national airspace system (NAS) was created at the dawn of commercial aviation to get aircraft from point A to point B in a safe and efficient manner.

(3) The NextGen program is meant to improve the current NAS by finding methods to deal with the increased traffic and improve the overall system.

(4) However, because the weather site is on the ground (and doesn't move with the plane), the pressure reported at the site may differ from the pressure at the plane's actual location, affecting the accuracy of the altimeter reading.

(5) Density altitude is important for determining the performance of an aircraft, or how the aircraft will behave under certain condition.

IV   Supplement Reading

Airspace Classification and Scope in Class D Airspace

In the U. S. , airspace consists of classes A, B, C, D, E, and G. The NAS includes both controlled and uncontrolled airspace.

Class A begins and includes 18,000 ft. MSL, and continues up to 60,000 ft. MSL. It is the most controlled airspace and requires a pilot to carry an Instrument Flight Rating and proper clearance no matter what type of aircraft is being flown. Pilots are also required to change their altimeter settings to 29.92 in. to ensure that all pilots within the airspace have the same reading in order to achieve proper altitude separation.

Class B airspace extends from the surface up to 10,000 ft. AGL and is the area above and around the busiest airports (e. g. , LAX, MIA, CVG) and is also heavily controlled. A side view of Class B airspace resembles an upside-down wedding cake with three layers becoming bigger toward the top. Class B's layers are designed individually to meet the need of the airport they overlay. Pilots must also receive clearance to enter the Class B airspace but Visual Flight Rules may be used, unlike in Class A airspace. Class B airspace corresponds to the area formerly known as a Terminal Control Area or TCA.

Class C airspace reaches from the surface to 4,000 ft. AGL above the airport which it surrounds. Class C airspace only exists over airports which have an operational control tower, are serviced by a radar approach control, and have a certain number of instrument flight operations. Class C is also individually designed for airport but usually covers a surface area of about 5 nautical miles around the airport up to 1,200 ft AGL. At 1,200 ft. the airspace extends to 10 nautical miles in diameter which continues to 4,000 ft. Pilots are required to establish two-way radio communications with the ATC facility providing air traffic control service to the area before entering the airspace. Within Class C, Visual and Instrument pilots are separated.

Class D airspace exists from the surface to 2,500 ft. AGL above an airport. Class D airspace only surrounds airports with an operational control tower. Class D airspace is also tailored to meet the need of the airport. Pilots are required to establish and maintain two-way radio communications with the ATC facility providing air traffic control services prior to entering the airspace. Pilots using Visual Flight Reference must be vigilant for traffic as there is no positive separation service in the airspace. This airspace roughly corresponds to the former Airport Traffic Area.

Class E airspace is the airspace that lies between Classes A, B, C, and D. Class E extends from either the surface or the roof of the underlying airspace and ends at the floor of the controlled airspace above. Class E exists for those planes transitioning from the terminal to en-route state. It also exists as an area for instrument pilots to remain under ATC control without flying in a controlled airspace. Under visual flight condition, Class E can be considered as uncontrolled airspace.

Airports without operational control towers are uncontrolled airfields. Pilots in these areas are responsible for position and separation and may use a specified Common Traffic Advisory Frequency (CTAF) or UNICOM for that airport, although no-radio flight is also permitted.

Class G airspace is uncontrolled airspace which extends from the surface to either 700 or 1,200 ft. AGL depending on the floor of the overlying Class E, or to the floor of Class A where there is no overlying Class E. In the vicinity of an uncontrolled airport, the CTAF for that airport is used for radio communication among pilots. In remote areas other frequencies such as MULTICOM are used. No towers or in-flight control services are provided although communications may be established with flight service stations which are not part of the NAS and advisory service may be available from ARTCC.

Class D airspace is the it space that surrounds airports that have an operating air traffic control tower, but don't have radar services (or at least the airport is not required to have radar). The airspace surrounding a Class D airport is only classified as a Class D airport when the air traffic control tower is in operation.

When an airport's traffic control tower is not in operation, the airport reverts from a Class D airport to a Class E airport or a combination of a Class E and Class G airport.

The control tower at a Class D airport provides traffic separation for Visual Flight Rules (VFR) and Instrument Flight Rules (IFR) traffic and can provide VFR traffic advisories, workload permitting.

How to Define Class D Airspace? The following factors need to be considered when defining Class D airspace.

- Dimension: Class D airports can be tailored to fit instrument approach procedures that are available at the airport, which means that one Class D airport might look slightly different from another. Literally, the dimensions depend upon instrument approach procedures into a particular airport. It's worth noting that there are extensions for arrivals and departures factored into the airspace dimensions. Normally, the airspace surrounding a Class D airport extends from the surface up to a designated MSL altitude, usually about 2,500 feet Above Ground Level (AGL).

- Entry Requirement: To enter Class D airspace, a pilot must contact the control tower and establish two-way radio communication prior to entering the airspace. During the call, you must provide Air Traffic Control (ATC) with your position, altitude, current transponder code, and your destination/intentions. You'll know you've established two-way communication when you get a call-back that includes your callsign (tail number). If you do not hear your number, you can not enter the airspace. If the controller is busy, they can ask you to stay out of Class D airspace until they are ready.

- Speed Restriction: In any airspace, aircraft can't exceed 250 knots when below 10,000 feet Mean Sea Level (MSL). However, when you're within 4 Nautical Miles (NM) of the primary Class D airport and at or below 2,500 feet AGL, you can't exceed 200 knots.

- Weather Minimum: Class D minimum weather requirements exist so that you can see and avoid other aircraft. Because not all Class D control towers have radar scope, ATC wants you to stay far enough away from the cloud so that you can see and avoid other airplane, especially the jet flying fast approaches. An easy way to remember VFR weather minimums for Class D airspace is the phrase "3 Cessna 152s." Each number in the phrase strands for a distance (i. e., 3 stands for 3 statical miles; 1 stands for 1,000 feet above the cloud; 5 stands for 500 feet below the cloud, and 2 stands for 2,000 feet away from horizontal cloud).

- VFR Visibility and Cloud Clearance Requirement: Pilots flying into Class D airspace must maintain at least three statute miles visibility. In addition, pilots must remain at least 500 feet below cloud, 1,000 feet above cloud, and stay 2,000 feet away from horizontal cloud while in Class D airspace.

- Chart Depiction: Class D airspace is depicted as a dashed blue line on a VFR sectional chart. You may notice that the airspace extends into the arrival and departure paths for IFR traffic.

Other Airspace Classifications

There are other classifications of controlled airspace (and their defined dimension) within which ATC service is provided. The other classifications include class A, B, C, E, and G.

- Class A Airspace: Class A airspace is generally airspace from the surface to 18,000 feet MSL up to and including Flight Level (FL)600, and the airspace overlying the waters within 12 NM of the coast of the 48 contiguous states and Alaska.

- Class B Airspace: Class B airspace is generally airspace from the surface to 10,000 feet MSL surrounding the nations' busiest airport in terms of airport operations or passenger capacity.

- Class C Airspace: Class C airspace is from the surface to 4,000 feet above the airport elevations (charted in MSL) surrounding those airports that have an operational control tower, are serviced by a radar approach control, and have a certain number of IFR operations or passenger capacity.

- Class E Airspace: Class E airspace is the controlled airspace not classified as Class A, B, C, or D airspace. A large amount of the airspace over the United States is designated as Class E airspace. Typically, Class E airspace extends up to, but not including, 18,000 feet MSL (the lower limit of Class A airspace). All airspace above FL 600 is Class E airspace.

- Class G Airspace: Class G airspace is uncontrolled airspace and is the portion of the airspace that has not been designated as Class A, B, C, D, or E. This airspace extends from the surface to the base of the overlying Class E airspace. Although ATC has no authority or responsibility to control air traffic, you should remember there are VFR minimums that apply to Class G airspace.

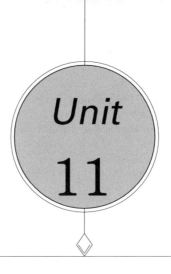

Unit 11

Flight Plan

I   Reading Text

Flight plan

Flight plans are documents filed by a pilot or flight dispatcher with the local Civil Aviation Authority (e. g. the FAA in the United States) prior to departure which indicate the plane's planned route or flight path. Flight plan format is specified in ICAO Doc 4444[Figure 1]. They generally include basic information such as departure and arrival points, estimated time en-route, alternate airports in case of bad weather, type of flight (whether instrument flight rules [IFR] or visual flight rules [VFR]), the pilot's information, number of people on board and information about the aircraft itself. In most countries, flight plans are required for flights under IFR, but may be optional for flying VFR unless crossing international borders. Flight plans are highly recommended, especially when flying over inhospitable area, such as water, as they provide a way of alerting rescuers if the flight is overdue. In the United States and Canada, when an aircraft is crossing the Air Defense Identification Zone(ADIZ), either an IFR or a special type of VFR flight plan called a DVFR (Defense VFR) flight plan must be filed. For IFR flight, flight plans are used by air traffic control to initiate tracking and routing services. For VFR flight, the only purpose is to provide needed information and search and rescue operations required, or for use by air traffic control when flying in a "Special Flight Rules Area".

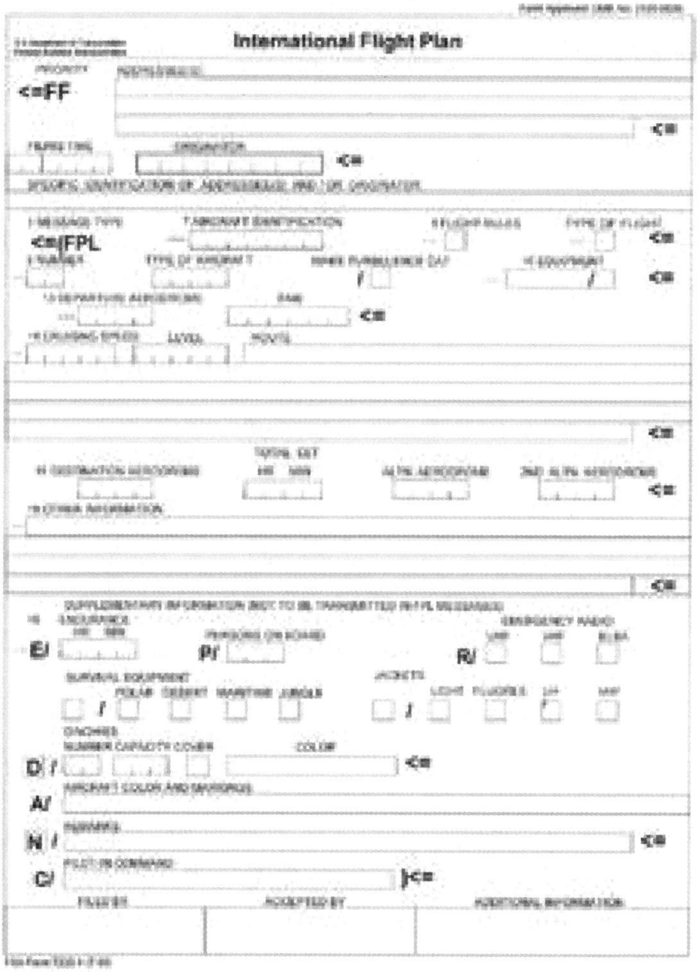
International Flight Plan

<≡FF

<≡(FPL

E/

P/

R/

D/

A/

N/

C/

Figure 1 International flight plan

Route or flight paths

Routing types used in flight planning are: airway, navaid and direct. A route may be composed of segments of different routing types. For example, a route from Chicago to Rome may include airway routing over the U. S. and Europe, but direct routing over the Atlantic Ocean.

Airway or flight path

Airway routing occurs along pre-defined pathways called flight paths. Airways can be thought of as three-dimensional highways for aircraft. In most land areas of the world, aircraft are required to fly airways between the departure and destination airports. The rules governing airway routing cover altitude, airspeed, and requirements for entering and leaving the airway (see SIDs and STARs). Most airways are eight nautical miles (14 kilometers) wide, and the airway flight levels keep aircraft separated by at least 1,000 vertical feet from aircraft on the flight level above and below. Airways usually intersect at Navaids, which designate the allowed points for changing from one airway to another. Airways have names consisting of one or more letters followed by one or more digits (e. g. , V484 or UA419).

The airway structure is divided into high and low altitudes. The low altitude airways in the U. S. which can be navigated using VOR Navaids have names that start with the letter V, and are therefore called Victor Airways. They cover altitudes from approximate 1,200 feet above ground level (AGL) to 17,999 feet (5,486 m) above mean sea level (MSL). T routes are low altitude RNAV only routes which may or may not utilize VOR NAVAIDS. The high altitude airways in the U. S. have names that start with the letter J and are called Jet Routes, or Q for Q routes. Q routes in the U. S. are RNAV only high altitude airways, whereas J routes use the same way as VOR NAVAID's which V routes do. J & Q routes run from 18,000 feet (5,486 m) to 45,000 feet (13, 716 m). The altitude separating the low and high airway structures varies from country to country. For example, it is 19,500 feet (5,944 m) in Switzerland, and 25,500 feet (7,772 m) in Egypt.

Navaid

Navaid routing occurs between Navaids (short for Navigational Aids, see VOR) which are not always connected by airways. Navaid routing is typically only allowed in the continental U. S. If a flight plan specifies Navaid routing between two Navaids which are connected via an airway, the rules for that particular airway must be followed as if the aircraft was flying Airway routing between those two Navaids. Allowable altitudes are covered in Flight Level.

Direct

Direct routing occurs when one or both of the route segment endpoints are at a latitude/longitude which is not located at a Navaid. Some flight planning organizations specify that checkpoints generated for a Direct route be a limited distance apart, or limited by time to fly between the checkpoints (i. e. direct checkpoints could be farther apart for a fast aircraft than for a slow one).

SIDs and STARs

SIDs and STARs are procedures and checkpoints used to enter and leave the airway system by aircraft operating on IFR flight plans. There is a defined transition point at which an airway and a SID or STAR intersect.

A SID, or Standard Instrument Departure, defines a pathway out of an airport and onto the airway structure. A SID is sometimes called a Departure Procedure (DP). SIDs are unique to the associated airport.

A STAR, or Standard Terminal Arrival Route ("Standard Instrument Arrival" in the UK) defines a pathway into an airport from the airway structure. STARs can be associated with more than one arrival airport, which can occur when two or more airports are in proximity (e. g. , San Francisco and San Jose).

Special use airspace

In general, flight planners are expected to avoid areas called Special Use Airspace (SUA) when planning a flight. In the United States, there are several types of SUA, including Restricted, Warning, Prohibited, Alert, and Military Operations Area (MOA). Examples of Special Use Airspace include a region around the White House in Washington, D. C. , and the country of Cuba. Government and military aircraft may have different requirements for particular SUA area, or may be able to acquire special clearances to traverse through these areas.

Flight levels

Flight levels (FL) are used by air traffic controllers to simplify the vertical separation of aircraft and one exists every 1,000 feet relative to an agreed pressure level. Above a transitional altitude, which varies from country to country, the worldwide arbitrary pressure datum of 1,013.25 millibar or the equivalent setting of 29.92 inches of mercury enter into the altimeter and altitude is then referred to as a flight level. The altimeter reading is converted to a flight level by removing the trailing two zeros: for example, 29,000 feet becomes FL290. When the pressure at sea level is by chance the international standard then the flight level is also the altitude. To avoid confusion, below the transition altitude, height is referred to as a numeric altitude, for example "descend 5,000 feet" and above the transition altitude, "climb flight level 250".

Airways have a set of associated standardized flight levels (sometimes called the "flight model") which must be used when on the airway. On a bi-directional airway, each direction has its own set of flight levels. A valid flight plan must include a legal flight level at which the aircraft will travel the airway. A change in airway may require a change in flight level.

In the USA and Canada, for eastbound (heading 0—179 degrees) IFR flights, the flight plan must list an "odd" flight level in 2,000 foot increments starting at FL190 (i. e., FL190, FL210, FL230, etc.); Westbound (heading 180—359 degrees) IFR flights must list an "even" flight level in 2,000 foot increments starting at FL180 (i. e., FL180, FL200, FL220, etc.). However, Air Traffic Control (ATC) may assign any flight level at any time if traffic situations merit a change in altitude.

Aircraft efficiency increases with height. Burning fuel decreases the weight of an aircraft which may then choose to increase its flight level to further improve fuel consumption. For example, an aircraft may be able to reach FL290 early in a flight, but step climb to FL370 later in the route after weight has decreased due to fuel burn off.

Alternate airports

Part of flight planning often involves the identification of one or more airports which can be flown to in case of unexpected conditions (such as weather) at the destination airport. The planning process must be careful to include only alternate airport which can be reached with the anticipated fuel load and total aircraft weight, and have capabilities necessary to handle the type of aircraft being flown.

In Canada, unlike the United States, unless specifically exempted by a company Operating Certificate, IFR flight plans require an alternate airport, regardless of the forecast destination weather. In order to be considered as a legally valid alternate, the airport must be forecast to be at or above certain weather minima at the estimated time of arrival (at the alternate). The minimum weather conditions vary based on the type of approach(es) available at the alternate airport, and may be found in the General section of the Canada Air Pilot (CAP).

Fuel

Aircraft manufacturers are responsible for generating flight performance data which flight planners use to estimate fuel need for a particular flight. The fuel burn rate is based on specific throttle settings for climbing and cruising. The planner uses the projected weather and aircraft weight as input to the flight performance data to estimate the necessary fuel to reach the destination. The fuel burn is usually given as the weight of the fuel (usually pounds or kilograms) instead of the volume (gallons or litres) because aircraft weight is critical.

In addition to standard fuel need, some organizations require that a flight plan include reserve fuel if certain conditions are met. For example, an over-water flight of longer than a specific duration may require the flight plan to include reserve fuel. The reserve fuel may be planned as extra which is left over on the aircraft at the destination, or it may be assumed to be burned during flight (perhaps due to unaccounted for differences between the actual aircraft and the flight performance data).

In case of an in-flight emergency it may be necessary to determine whether it is quicker to divert to the alternate airfield or continue to the destination.

Flight plan timeline

Flight plans may be submitted before departure or even after the aircraft is in the air. However, flight plans may be submitted up to 24 hours in advance either by voice or by data link; though they are usually filled out or submitted just several hours before departure. The minimum recommended time is 1 hour before departure for domestic flight, and up to three hours before international flight. This time depends on the country the aircraft is flying out of.

Other flight planning consideration

Holding over the destination or alternate airports is a required part of some flight plans. Holding (circling in a pattern designated by the airport control tower) may be necessary if unexpected weather or congestion occurs at the airport. If the flight plan calls for hold planning, the additional fuel and hold time should appear on the flight plan.

Organized Tracks are a series of paths similar to airways which cross ocean areas. Some organized track systems are fixed and appear on navigational chart (e. g. , the NOPAC tracks over the Northern Pacific Ocean). Others change on a daily basis depending on weather, west or eastbound and other factors and therefore cannot appear on printed charts (e. g. , the North Atlantic Tracks (NAT) over the Atlantic Ocean).

Description of flight plan blocks (FAA) and Domestic Flight Plan Form 7233 – 1

Standard FAA flight plan form includes [Figure 2]:

1. Type: Type of flight plan. Flights may be VFR, IFR, DVFR, or a combination of types, termed composite.

2. Aircraft Identification: The registration of the aircraft, usually the flight or tail number.

3. Aircraft Type/Special Equipment: The type of aircraft and how it's equipped. For example, a Mitsubishi Mu-2 equipped with an altitude reporting transponder and GPS would use MU2/G. Equipment codes may be found in the FAA Airman's Information Manual.

4. True airspeed in knots: The planned cruise, true airspeed of the aircraft in knots.

5. Departure Point: Usually the identifier of the airport from which the aircraft is departing.

6. Departure Time: Proposed and actual times of departure. Times are Universal Time Coordinated.

7. Cruising Altitude: The planned cruising altitude or flight level.

8. Route: Proposed route of flight. The route can be made up of airways, intersections, navaids, or possibly direct.

Figure 2　Flight plan

9. Destination: Point of intended landing. Typically the identifier of the destination airport.

10. Estimated Time Enroute: Planned elapsed time between departure and arrival at the destination.

11. Remarks: Any information the PIC believes is necessary to be provided to ATC. One common remark is "SSNO", which means the PIC is unable or unwilling to accept a SID or STAR on an IFR flight.

12. Fuel on Board: The amount of fuel on board the aircraft, in hours and minutes of flight time.

13. Alternate Airports: Airports of intended landing as an alternate of the destination airport. May be required for an IFR flight plan if poor weather is forecast at the planned destination.

14. Pilot's Information: Contact information of the pilot for search and rescue purpose.

15. Number Onboard: Total number of people on the aircraft.

16. Color of Aircraft: The color helps identify the aircraft to search and rescue personnel.

17. Contact Information at Destination: Having a means of contacting the pilot is useful for tracking down an aircraft that has failed to close its flight plan and is possibly overdue or in distress.

Ⅱ Words and Expressions

dispatcher	*n.* 调度员,签派员
specified	*adj.* 明确要求;具体说明
overdue	*adv.* 过期未完成的;过期未付的;早该发生的;期待已久的
initiate	*vt.* 开始实施,发起;使认识;传授
rescue	*vt.* 营救,解救,救援
governing	*adj.* 统治的,执政的,管理的
designate	*vt.* 任命,选定,指派
utilize	*vt.* 利用,使用
continental	*adj.* 大陆性的;欧陆风格的
segment	*n.* 部分,段,片
endpoint	*n.* 定义统计分析的结束点;端点,终点,端点绘制
intersect	*vt.* 相交,交叉;分隔
associated	*adj.* 与……相关联的
proximity	*vt.* 接近,临近;邻近
arbitrary	*adj.* 任意的;专横的,武断的
Mercury symbol Hg	*n.* 汞,符号为 Hg
eastbound	*adj.* 向东行的,朝东驶的
increment	*n.* 增加量;定期加薪
merit	*n.* 长处,优点
alternate	*adj.* 间隔的;轮流的,相间的,交替的;可替代的
anticipated	*adj.* 受期盼的
valid	*adj.* 有效的;正式认可的;合理的理由等;系统认可的
throttle	*vt.* 掐,勒;掐死;阻挡;压制;扼杀
volume	*n.* 音量,响度;流量,总量
formula	*n.* 准则;方案;公式;方程式;分子式

groundspeed   *n.* 地速

remark   *vt.* 说；议论；评论

Ⅲ   Exercises

1   Translate the following phrases into English

(1) 防空识别区

(2) 军事活动区域

(3) 标准仪表离场程序

(4) 飞行高度层

(5) 标准进场程序

(6) 平均海平面

(7) 过渡高度

(8) 改航去备降

(9) 最低天气条件

(10) 阶段爬升

2   Translate the following sentences into Chinese

(1) The planning process must be careful to include only alternate airport which can be reached with the anticipated fuel load and total aircraft weight, and have capabilities necessary to handle the type of aircraft being flown.

(2) The reserve fuel may be planned as extra which is left over on the aircraft at the destination, or it may be assumed to be burning during flight (perhaps due to unaccounted for

differences between the actual aircraft and the flight performance data).

(3) Airways have a set of associated standardized flight levels (sometimes called the "flight model") which must be used when on the airway.

(4) The minimum recommended time is 1 hour before departure for domestic flight, and up to three hours before international flight.

(5) The planner uses the projected weather and aircraft weight as input to the flight performance data to estimate the necessary fuel to reach the destination.

IV   Supplement Reading

Smooth Aircraft Approach Cuts Noise Pollution

By Paul Marks

Noise nuisance from aircraft can be reduced significantly by changing the way the planes come in to land. Lining up with the runway as far as 70 kilometres away and making a steady descent can more than halve the acoustic energy that reaches the ground, an international research consortium has found.

Now pressure is on for the technique to be considered for the busiest international airport in the world, Heathrow, in London, UK.

Noise pollution around airports is set to get worse. Air traffic worldwide is increasing at 4.7 per cent per year and is expected to triple by 2030. Without a major initiative to reduce aircraft noise, airports will be prevented from handling the extra flight, says Mike Howse, director of technology and engineering at the British aero-engine maker Rolls-Royce.

The noise during descent comes from two sources [Figure 3]: the engine, particularly when they have to deliver high power as the plane manoeuvres near to the ground, and aerodynamic noise from the flap on the trailing edge of the wing, which also gets worse during manoeuvres.

So the way to reduce noise near the ground seems straightforward: carry out the manoeuvres needed to get on course for the runway farther away from airports and at high altitude, and then descend slowly with minimal correction.

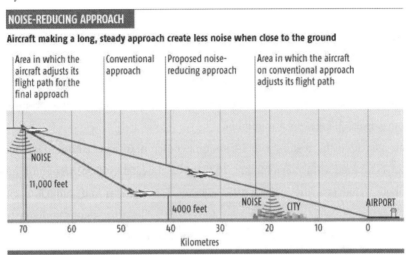

Figure 3   Noise Reducing Approach

Continuous descent

A method of noise reduction along these lines, known as a continuous descent approach, has already been in use at many airports. But now a consortium that includes Rolls-Royce, the Massachusetts Institute of Technology, the University of Cambridge, British Airways and the UK Civil Aviation Authority wants to take the idea further.

In a continuous descent approach, an aircraft begins its final descent from a distance of about 17 kilometres and an altitude of 4,000 feet. It then maintains a steady 3° angle of descent during its approach, says Richard Wright, spokesman for the UK's National Air Traffic Services.

But this technique involves in significant manoeuvring at 4,000 feet and below, which can be noisy for people living under that part of the flight path. "What we need is an approach procedure where there are no transient in thrust over noise-sensitive locations," says John-Paul Clarke, an aeronautical engineer at MIT.

Using a flight simulator modified to analyse noise, he and colleagues at Boeing and NASA developed a quieter continuous descent path into Louisville International Airport in Kentucky. Working with the cargo airline UPS, Clarke then asked the crew of two Boeing 767 jets to adopt the new approach routine.

Acoustic energy

Instead of beginning the descent 17 kilometres away, the crew began their descents some 70 kilometres away and from a height of 11,000 feet, while maintaining an angle of descent of 3° and keeping engine power changes to a minimum.

The team measured the results with 14 microphones placed in neighbourhood near the airport and found that compared with conventional 767 approaches, the noise dropped by between 3.9 and 6.5 decibels. That is a drop in acoustic energy of more than half, and is noticeable to the human ear.

What's more, the manoeuvre saved nearly 200 kilograms of fuel on each landing, Clarke says. The team has submitted its research to the Journal of Aircraft.

Complex patterns

London is served by five airports whose in and outbound air lanes form complex patterns above the city and require aircraft to manoeuvre at relatively low altitude. Wright says it could be extremely difficult to reorganise the airspace in a way that does not require low-altitude manoeuvres.

However, the public may force the issue. John Stewart of the pressure group Heathrow Association for the Control of Aircraft Noise welcomes the idea. "These experiments are worth doing, whatever the challenges," he says.

"It certainly sounds like there is an upside for people living near the airport. We'd need to know what it means for people further out. But the technique should be considered for Heathrow."

Unit 12

Visual Flight Rule and Instrument Flight Rule

Ⅰ Reading Text

VMC Vs. VFR

By Michael Vivion and Jeff Berlin

In basic flight training, student pilots memorize the cloud clearance and visibility criteria for operation under visual flight rule and instrument flight rule (VFR and IFR).

Flight schools and instructors drill into students the cloud clearance and visibility requirements for VFR operations in various categories of airspace, all the while neglecting to mention that none of this has much to do with the ability to keep an airplane upright during periods of restricted visibility and/or lack of terrain definition.

As a matter of fact, aside from concerns about FAA violations, what pilots should really care about is whether they have sufficient visual references to keep an airplane right-side up and navigate visually. Those are the things that keep us alive in VFR flight, and in comparison, an FAA violation is no big deal. The condition I'm describing are VMC, or visual meteorological condition, as compared to VFR.

It's essential that pilots understand the operational difference between VFR and VMC. VFRs are precisely what the name implies: a set of visibility and cloud-clearance standards that dictate the type of flight operation (VFR or IFR)that a pilot may legally conduct.

VMC, on the other hand, refers to conditions that permit a pilot to maintain visual reference with a horizon and provide enough visual references by which he or she can navigate. Pilots should clearly understand what constitutes VMC and commit to a strategic 180-degree turn in the event that they're no longer able to navigate with visual reference. Another alternative is a fast and effective transition to instrument flight upon entry into instrument meteorological condition (IMC), even though the visibility and cloud-clearance requirements meet the basic VFR standard.

Unfortunately, as the NTSB database illustrates, pilots don't do a very good job of making these transitions in real life. Having operated in IMC but under VFR on several occasions, I can testify that it's a difficult transition to make.

Confused yet? Let me provide a couple of examples of what I refer to as "PVFR."

Marginal VFR condition can very quickly transition into IMC[Figure 1].

It's winter in northwest Alaska. The forecast for a flight eastbound from Kotzebue calls for VFR condition along the route. Current visibilities are at least five miles, and ceilings are 3,000 feet or greater at all reporting stations along the route. It looks like a go.

But first I need to verify that the instrumentation in the aircraft is functional and that the airplane is approved for instrument flight. As the pilot, I also need to be current and competent (not necessarily the same thing, as the NTSB accident database graphically illustrates) on instrument. At least the first portion of the flight I'm about to embark on will be through weather condition that I refer to as PVFR. If you haven't found that acronym in the Aeronautical Information Manual, it's because I made it up.

Figure 1 Marginal VFR

My definition of PVFR is: "A meteorological condition in which visibility and ceilings meet

the minimum required value specified in the Federal Aviation Regulations for flight under VFR...
but there's nothing within the prevailing visibility to use as a visual reference." The P in PVFR
stands for "pretend," in case you haven't guessed by now.

The snow-covered and treeless landscape that I'm preparing to traverse provides no visual
definition, and the low winter-sun angle produces what's referred to as "flat light." The
combination of flat light and absence of visual references produce a phenomenon described by
many pilots as flying inside a milk bottle.

The condition for this flight meet the requirement for a basic VFR operation, but in fact,
IMC will prevail along a significant portion of the route, and the only way to safely conduct this
flight is by reference to instruments.

Few pilots experience the joys of winter flight in northern Alaska, and those who do quickly
learn to either stay on the ground during marginal condition or to exercise instrument flying skills
to get from point A to point B, but PVFR isn't exclusive to Alaska.

On a clear, dark night over the desert or other unpopulated area, what's technically VMC
can become dangerous to the VFR-only/non-instrument-rated pilot.

Consider the East Coast of the United States on a hot summer afternoon. Humidity rivals that
of a steam room, and temperatures soar. Haze builds until three miles seems like pretty good
visibility. In contrast to winter in the north, there are plenty of visual reference here, so keeping
the airplane right-side up isn't difficult at all. Or is it?

What if your flight follows the Atlantic coast to a coastal village? The last leg of the flight will
take you over a portion of the Atlantic Ocean. Visibilities at reporting station along the coast are
all about five miles in haze. Fairly good basic VFR, right?

The flight includes an intermediate stop to pick up a passenger. Delays caused by business
and traffic en-route to the airport cause you to arrive at your intermediate stop just after sunset.
You spend a few minutes loading your passengers, and then you launch toward your weekend
destination. This is the leg of the trip that crosses water, but it's a beautiful evening with calm
wind and smooth sea surface.

The route of flight extends a maximum of about seven miles from shore at one point, crossing
a large bay. As you proceed toward your destination, you realize that there's no longer a visual
horizon. The ocean surface is impossible to see as well. Lights on shore are seven miles away—in
five-mile visibility.

John F. Kennedy Jr. and his passengers lost their lives in similar condition a few years ago.

This scenario may be a little closer to home than the Alaska flight described earlier, but the conditions are very much PVFR.

Finally, consider a flight in low but legal VFR condition near Oshkosh, Wis. Departing from the big air show in four-mile visibility, you're homeward bound—to the east. Reporting stations to the east are declaring clear skies and visibility of 10 miles. To the east, however, lies Lake Winnebago. Do you go around the north end of the big lake or go direct? Again visual reference may be totally absent over the lake, even though visibility is above basic VFR.

Perhaps the most significant difficulty in these situations is for the pilot to recognize that he or she is entering IMC, even though the visibility and ceilings meet the basic VFR requirement. Operating visually in PVFR condition can rapidly result in a loss of control, unless the pilot quickly recognizes the deteriorating condition and performs a 180-degree turn to exit those condition, or successfully and quickly transitions to flight by instrument.

So what can we do to better prepare ourselves for a possible foray into PVFR? The answers to this question are simple, but the action implied by these answers are more difficult to accomplish than they may seem at first blush.

Figure 2   Preflight weather briefing

When acquiring your preflight weather briefing, you should investigate weather pattern in areas you frequent (be more conservative in less-familiar destination). Additionally, pay attention to the landscape and potential obstacles you could encounter on your next flight[Figure 2].

First, pilots should view visibility forecast with a skeptical eye. I don't mean to suggest that forecasts are always flawed. I'm thankful that I'm not a forecaster since accurate forecasting, particularly over large parts of the country, is a daunting task. There's no doubt in my mind that forecasters do the best they can, but the forecasts they generate are frequently in error. So when acquiring your preflight weather briefing, dig a bit deeper than just the basics. Try to discover the trend. Find out what direction the weather is moving and look at observed weather condition "upstream" from your route of flight. Consider whether the observed weather is better or worse than the previous forecast suggested it should be. Learn about the weather pattern in areas you frequent and be more conservative in areas you're unfamiliar with.

Next, look at charts of the area. Are there portions of your route that may offer poor visual reference? Crossings of large lake or the extensive prairie country in the Midwest in winter can create a PVFR scenario in low ceiling or restricted visibility.

Talk to the local. I've found that most pilots are happy to provide visitors with advice on flying safely in their part of the country. Warren Thompson in Kotzebue was responsible for my introduction to PVFR in that part of the world. Thompson worked for many years at the Kotzebue Flight Service Station, and in his free time, he flew for an air-taxi operator. Thompson not only briefed the forecasts to pilots in the area, but also flew the weather himself. Unfortunately, the consolidation of Flight Service Station has nearly eliminated this kind of local knowledge among FSS briefers.

Be conservative. This is easy to say while you're at home, perusing the weather via the Internet or talking to a briefer. Spend three days camping in the rain at a small airport a few hundred miles from home and that conservatism gets harder to maintain.

If you climb into an airplane and launch telling yourself you can always turn around, it's even harder to fold your card and perform a 180-degree turn, but that critical maneuver must always remain as an option.

Lastly, maintain instrument proficiency. I offer this as a last resort because, as noted, the accident database suggests that even highly experienced, instrument-rated pilots often do a terrible job of making the transition from VMC to instrument flight. Climbing into IMC and transitioning to instrument flight from the visual should be a planned act, not a get-out-of-jail-free card, to be played only if things really turn sour on you.

Most of all, every pilot should understand that even basic VFR weather can create condition under which it's impossible to operate an aircraft by visual reference. As you plan your flight, consider not only the weather condition you're likely to encounter, but also the terrain and any obstacles you're likely to encounter en-route, and whether or not you can maintain visual flight under those conditions.

II　Words and Expressions

clearance	*n.* 许可,批准;清除,清理
criteria	*n.* 标准
neglect	*vt.* 疏于照管;忽视
meteorological	*adj.* 气象学的
essential	*adj.* 极其重要的;必不可少的;最基本的;本质的
dictate	*vt.* 口授,让听写;命令,规定,指定;支配,影响,决定
constitute	*vt.* 被视为,被算作;组成,构成;成立,设立
strategic	*adj.* 战略性的;策略上的;有用的,合适的;战略性的,战略上的
illustrate	*vt.* 说明,阐明
competent	*adj.* 合格的,能干的,胜任的;合格的,过得去的
embark	*vt.* 上飞机;装飞机
portion	*n.* 部分;一份;命运
	vt. 分配;给……嫁妆
acronym	*n.* 首字母缩略词
traverse	*vt.* 跨过,穿过,横越,横穿
prevail	*vi.* 盛行,流行;获胜
marginal	*adj.* 微小的,微不足道的,不重要的;边际的
exclusive	*adj.* 独有的;独享的;专用的;高级的;昂贵的
humidity	*n.* 湿度;湿气
haze	*n.* 烟雾,霾;懵懂,迷糊,迷惑
intermediate	*adj.* 中级的,中等的;中间的,居中的
scenario	*n.* 方案;情节;设想
deteriorate	*vt.* 恶化
foray	*n.* 短暂尝试;突袭
blush	*vi.* 脸红;感到羞愧;尴尬
skeptical	*adj.* 持怀疑态度的,不相信的

flawed	*adj.* 有错误的;有缺点的;有损坏的
daunting	*adj.* 使人气馁的,吓人的
upstream	*adv.* 向上游;逆流地
prairie	*n.* 大草原;牧场
eliminate	*vt.* 被淘汰;消除;排除
perusing	*vt.* 阅读
maneuver	*vt.* 熟练的动作;策略,巧计,花招
proficiency	*n.* 熟练;精通
encounter	*vt.* 遇到,遭遇;偶然碰到

Ⅲ Exercises

1 Translate the following phrases into English

(1)空域类型

(2)缺乏地表特征

(3)目视飞行规则的边界条件

(4)航空资料手册

(5)复飞

(6)联邦航空法规

(7)失控

(8)天气模式

(9)转弯

(10)目视参考

2 Translate the following sentences into Chinese

(1)As a matter of fact, aside from concerns about FAA violations, what pilots should really care about is whether they have sufficient visual reference to keep an airplane right-side up and navigate visually.

(2) Few pilots experience the joys of winter flight in northern Alaska, and those who do quickly learn to either stay on the ground during marginal condition or to exercise instrument flying skills to get from point A to point B.

(3) Perhaps the most significant difficulty in these situations is for the pilot to recognize that he or she is entering IMC, even though the visibility and ceilings meet the basic VFR requirement.

(4) If you climb into an airplane and launch telling yourself you can always turn around, it's even harder to fold your card and perform a 180-degree turn, but that critical maneuver must always remain as an option.

(5) As you plan your flight, consider not only the weather condition you're likely to encounter, but also the terrain and any obstacles you're likely to encounter en-route, and whether or not you can maintain visual flight under those conditions.

IV   Supplement Reading

How to Become an Instrument-Rated Pilot and Illusions Pilots Encounter While Flying

Instrument Training Hones Your Skills

By Sarina Houston

An instrument pilot can legally fly in the cloud, rain, and fog, which broadens his abilities and keeps him in the air instead of on the ground during inclement weather. Pilots who want to fly in the cloud need to to get an instrument rating added on to their private or commercial pilot certificate. And most professional aviation businesses require pilots to be instrument – rated, so it's a necessary step for those who might want to become an airline pilot or corporate pilot. The ability to fly solely by reference to instruments in the aircraft means that a pilot isn't limited to good weather operation only.

An applicant for an instrument rating needs to be extremely precise and detail-oriented. He or she must be able to follow procedures and multitask to a higher level than before. Since flying in inclement weather with no visual reference to the ground can be hazardous for an untrained pilot, instrument training requires a great deal of professionalism and leaves no room for mistakes or carelessness.

If you're trained well and take it seriously, IFR flying can be very rewarding. Here's what you'll need to do to become an instrument rated pilot.

- Know the Eligibility Requirement

Instrument pilot applicants must be able to read, speak, write and understand English, and must hold at least a private pilot certificate.

- Study For and Take the FAA Written Exam

Just like with previous pilot certificates you may have earned, you'll want to get the written exam out of the way early in your instrument training. That way you'll have the extra knowledge in your head ahead of time, or it will also serve as a refresher if you've taken some time off. Once your written exam is completed, you can focus on flying.

- Invest in Some New Pilot Supplies

You'll need to be more organized than ever in the cockpit, so get some new IFR supplies that work for you. Many IFR pilots like using iPads or similar devices for task management. You may also need a binder for your charts, a timer, and "foggles" (fogged-up goggles that simulate IFR flight during good weather training flight). If you have the money to spend, you may also want to consider a handheld GPS device as a backup to any onboard equipment your aircraft might have. Handheld units are not IFR-certified, but would come in handy during an emergency or if you happen to lose situational awareness on an actual IFR flight. (Keep in mind that your instructor and check pilot may not allow these during training.)

- Start Flying

To obtain an instrument rating under CFR Part 61, you'll need at least 50 hours of pilot-in-command (PIC) cross-country flight time. You'll need 40 hours of actual or simulated instrument time, including at least one IFR cross-country flight that exceeds 250 nautical miles and include at least three different kinds of instrument approaches (one at each airport).

- Learn the Procedure

During your instrument training, you'll be assessed on procedures such as approaches, departures, holding, tracking, and intercepting courses. You'll practice emergencies during IFR condition and learn the ins and outs of navigational equipment. Most importantly, you'll learn situational awareness at a much higher level than earlier training. Typically, you'll perform a few cross-country flights to get adjusted to the real-world IFR environment-the "system" in which IFR pilots are handled.

- Take the Checkride

When you've mastered instrument flight and learned all about the privileges and limitations of the instrument rating, your instructor will sign you off for the checkride. Since you've taken checkrides before, you know what to expect: a couple of hours of ground work for the oral portion of the exam and another hour or two in the air to fly a few approaches is all it takes. For the instrument checkride, you'll have to fly at least two non-precision approaches and at least one precision approach. (Remember a GPS approach is a NON-precision approach!) One of these approaches will likely be a partial panel approach, in which you'll simulate an approach with failed instrument.

Remember the examiner is testing your ability to fly safely in low or zero visibility. In addition to knowing all of the procedures and tasks, you'll need to be extremely precise and always know exactly where you are!

For pilots, flying at night is mostly a pleasant task. It's often quiet as the radio chatter dies down for the day and smooth as the convective turbulence settles. But night flight also comes with its own set of challenges, including nighttime illusions. Pilots are trained to identify these illusions and ignore them or compensate for them while flying, but the night sky can be deceptive to even the best of pilots.

Here are nine types of illusions that pilots are confronted with.

The Black Hole Approach   The black hole approach occurs during an approach over a large, unlit area. Often it happens over bodies of water, but it can occur over any unlit terrain. With no visual reference over a large black hole, a pilot can easily overshoot or undershoot his position on the approach, resulting in an unstable approach. When experiencing a black hole approach illusion, a pilot should rely on the aircraft's instrument, remain at an appropriate altitude and work to maintain a stable approach, including a stable airspeed and descent rate.

Autokinesis   Autokinesis is an illusion of the eye. At night, when a pilot's eye stares at a light against a dark background with no other visual reference around it, such as a star or the light from another aircraft, the pilot will get the impression that the light is moving. Just knowing about this illusion helps demystify it, and moving the eyes around or looking to the side of a lit object can help.

False Horizons   VFR pilots rely heavily on the natural horizon of the Earth to maintain straight and level flight during the day. At night, when the sun goes down and there's no horizon to look at, the mind will often try to seek unsuccessfully one out. Often times, a pilot will interpret a misshapen cloud or the lights of a highway as a horizon and will bank the aircraft so that it's attitude is straight and level in relation to the newfound false horizon.

This is trouble, of course, as the result that in this case it is an unwanted consistent turn. A pilot flying at night needs to rely heavily on the attitude indicator in the aircraft to make sure he remains straight and level while recognizing such false horizons.

Flicker Vertigo   Flicker vertigo is a rare condition in which the brain doesn't process flickering light very accurately. It can be caused by the strobe light at night flickering against the propeller or from the sunlight on the propeller, and it results in disorientation and nausea. The good news is that along with its rare occurrence, it's fairly easy to rectify the pilot should just turn off the light or turn away from the sun.

Runway Lights   Bright runway lights can cause the pilot to feel like what the aircraft is lower than it really is, creating a situation in which a higher approach than normal is flown to compensate for what they feel is a high approach. An unstable approach condition is likely to occur if the pilot doesn't trust his instrument in this case.

Sloping Terrain   When the terrain slopes upward just before the approach end of the runway, the pilot can be tricked into believing the aircraft is too high, causing him to compensate by flying lower.

Conversely, a downhill slope will cause a pilot to think he is too low, resulting in a higher glide path than the normal .

Runway Width   A wider-than-usual runway will cause a pilot to think he is low. In an effort to compensate, he might fly a higher-than-normal approach, or pitch up to an unsafe airspeed on final approach.

Precipitation   Rain, fog, and haze can all cause pilots to perceive distance inaccurately. Rain, for example, can cause approach and runway lights to seem brighter at night, causing a pilot to feel like what he's lower than he should be, in which case he might overcorrect to a higher-than-normal approach. And fog and haze can both make the runway look farther away than it really is, causing an illusion of being too high.

White-Out Condition Snow-covered terrain combined with a gray overcast layer can cause a complete white – out illusion that makes it difficult for a pilot to gain any sort of visual reference, which means it's tough for a pilot to determine how high or low he is on the approach.

Paying close attention to the approach altitude and airspeed will help correct for this.

Illusions can cause disorientation in pilots, especially at night or in low-visibility condition. For almost all of these illusions, the fix is simple: Trust the instrument, maintain a stable approach speed and appropriate altitudes for the approach segment being flown, and be mentally prepared to recognize an illusion when it's occurring.

Unit 13

Human Factors

I   Reading Text

ICAO: Human Factors in Air Traffic Control

ICAO addressed the topic of human factors in Air Traffic Control (ATC) in the 1993 circular Human Factors Digest No. 8. Although a relatively old publication, the main ideas that are addressed in the circular remain relevant in the 21st century and described below is a synopsis of the contents in the circular.

Introduction

This digest focuses on human factor issues pertaining to ATC with the aim of articulating mainly to the ATC fraternity and to the larger aviation community the way forward for human factors vis-a-vis automation in ATC; Human Factors within Systems are introduced, whilst demonstrating how ATC performance and safety could be compromised due to inherent limitation in human.

The Evolution of ATC

Human Factors within Systems

An ATC system is a model of a human-machine system with the aim of achieving a safe and efficient flow of air traffic; a necessary pre-requisite for this being the continued supply of professional controllers (humans) well versed in the interaction and usage of technology (machines) available.

This interaction/matching of the human-machine system is an ongoing and changing process, hence it is imperative that humans are matched with systems successfully with the correct application of Human Factors data available so that the full benefit of the ATC system can be harnessed.

These Human Factors knowledge must thus be applied to both the effects of the human on the system and effects of the system on the human with the aim of improving safety and prevention of accidents.

Controllers also must have a firm grasp on how the system operates whilst utilizing his own professional attributes for the above to be achieved.

The shell model will be applied extensively to the ATC system in the digest and illustrates the main interactions of humans (Liveware) with other aspects of the system such as machines (Hardware), materials/procedures (Software), Environment and with other humans (Liveware).

However, it is important to note that Human Factor topics are always interlinked and must be dealt through its various interactions with each other instead of independent entities.

Evolution of ATC

Air Traffic came into prominence with the forming of ICAO in 1947 and the subsequent establishment of Rules of the Air and Air Traffic Services at a period of burgeoning demand for air travel and advances in technology both in aircraft and ATC equipment.

Airspaces round the world were demarcated into flight information regions (FIRs) under different international states; FIRs were further divided into classes of controlled and uncontrolled airspace (Class A—E) stipulating which type of ATC services were provided.

Three main categories of air traffic control (Area, Approach and Aerodrome) were identified specifying ATC responsibilities of aircraft from taxi to take off, en-route in the area, arrival phase and landing.

For the above to work properly, a flight plan must be submitted by the pilot with the detail containing identity, aircraft type and destination. Using these details which are displayed on the radar screen, controllers can de-conflict the aircraft (either vertically or horizontally) through the means of mandatory radar control via radio communication.

Future of ATC

There is now an exponential increase in air traffic around the world with ATC systems operating at close to near capacity for longer period of time.

With the distinct possibility of these capacities being exceeded, new avenues must be explored to deal with these increased demand; this includes increased automation, availability of better up to date information to the controller and changing existing mental models of reacting to new traffic scenarios to better advanced forward planning of efficient traffic flow.

The lack of an increase of controllers could also exacerbate the increasing demand to the limited airspace and present restrictions which might require more controllers to intervene a potential confliction should emerge.

In spite of all these possible changes and developments through the improvement in radar, collision avoidance and navigation systems must be balanced with the capabilities of humans with safety in ATC never compromised.

Transfer of Information

Efficiency of flight in ATC is dependent on several factors, such as characteristics of aircraft, equipment, how the aircraft is controlled, professionalism of ATC/pilot, information, the number of aircraft and environmental factor.

With experience, controllers are able to identify and react to poor qualitative (based on performance) and quantitative (accuracy of data) information about aircraft.

In most instances, it is the qualitative automation that determines how close aircraft can be from one another, hence increasing efficiency in the ATC system.

The Controller's Workspace

Application of Ergonomic Data

Controlling in ATC is dependent on the interaction of the human-machine interface and it is paramount that the workspace be set up with all of constraints resolved to facilitate maximum task performance through the application of ergonomics.

The broadest application of ergonomics to ATC workspace is environmental and examples include:

- Buildings must be designed with basic amenities closed by.
- Rooms must be spacious enough to accommodate all staff members comfortably.
- Controller's workstation must be equipped with all the necessary tools for controlling and if there is a need for sharing, information must be displayed clearly and within reach for both controllers. If various control positions collapse due to lull traffic, supervisors at their workstation must still be in the position to monitor all the controllers.
- In the aerodrome environment, there must be unimpeded visual access of ground vehicles and aircraft at critical phases of flight. There must also be a simple and seamless flow of information so that aircraft details are always up to date and available for handover to the next control agency.
- The controller's console must meet the ergonomic requirement for all body types of controllers with easy access to critical and frequently used control placed within eye distance.
- Anthropometry in the environment whereby the controller is working at his most comfortable position with ample space through the adjusting of his workspace must also be factored in.
- The layout of equipment must fulfill its roles and responsibilities and not placed for the sake of being there and should be configured to minimize and prevent distractions.
- Lighting in the form of ambient light and glare must be taken into account in a radar or aerodrome environment so that all the display and control remain visible for controlling without any distraction.
- The thermal environment must be maintained at a comfortable level through sound management of temperature and air flow rate for controlling.

- Noise levels within workstation should be kept at a minimum unless it becomes a distraction, causing important information and messages to be missed out by controllers.

- Visual display containing information for controllers must factor in human capabilities in how they see and process the information for it to be of good use. Hence, visual display must be clear and visible with information portrayed in the correct color contrast, color differentiated to avoid ambiguity if need be, with symbols and alphanumeric having the appropriate spacing for easy deciphering.

- Input devices (keyboard, mouse) in which the controller executes all control action must also be easy to be used with the necessary response sensitivity and feedback.

- An essential ingredient of ATC is communication and controllers must be aware of its uses and whether the system is functioning properly in his transmission out to the pilot. Standard radio telephony should be articulated slowly and clearly at all times to prevent any ambiguity and controllers must always verify if there is any doubt in the content of transmission.

Controller Proficiency

ATC involves a large amount of multi-tasking in which controllers must utilize all the different information available and draw on his past experience and memory to make the right decisions at all times. Hence human factors addresses the thinking process of controllers and the effects of equipment change on them with the aim of not drastically changing his thinking process, although in minor instances, retraining has to be done so that he can unlearn old habit and learn new technique.

Classes of Information

Controllers must have a thorough understanding of all the form of information available to them through the myriad of sources; it includes those from speech transmission and displayed information on the radar screen. Future ATC system will attempt to help controllers through more concise and processed information, alerting controllers to anomalies and conflictions and help in predicated future problem by offering solution of projected confliction in flight path. Notwithstanding the development of these sophisticated systems, controllers must still be conversant of all available information in the event of system contingencies, hence all information needed by the controller must be reliable and up to date.

Automation in ATC

Full or Partial Automation

Human Factors implication related to the liveware-software interface concerning partial automation intended to assist controllers will be delved into due to the problem associated with it.

Reason for Automation

With technological advance and an increasing amount of available information, automation is necessary to aid the controller so that the essential process in information management and safety in ATC are not compromised.

Goal of Automation

Automation could be of great benefit and an asset to control as it could bring about improvement in efficiency, safety, error prevention and reliability. Hence, it must correctly be matched with human capabilities through the maximizing of attributes and potential of both human and machine.

In spite of this, humans will still be the central element of the human-machine system with the machine helping the human and not vice versa. Hence, it is imperative that the human-machine system be studied carefully from an infant stage so that it is not made redundant and doesn't jeopardize the ATC system.

Constraint

Some of the constraints that need to be dealt with and are brought about by human function within an automated ATC system include:

- Human must not become over-reliant on automation so much that they aren't able to handle situation arising from system failure. Hence, human expertise level must be kept up to standard through regular currency.
- Controllers must remain plugged in with the whole ATC control process and not allow his awareness of traffic to be diminished by taking automation for granted.

- There must be a constant balance of workload between a minimum and maximum threshold so that safety is not compromised.

- The differing nature of workload involves non-interchangeable skill which may still need human to verify certain automated function.

- With automation, there is a distinct possibility that job satisfaction for human may be eroded.

- As part of a controller's proficiency, they must be cognizant of the benefit and limitation of the system so that they can trust and use it appropriately.

- Controllers must be clear of the division of task responsibilities between himself, other controllers and the automated system to prevent confusion.

- The co-ordination process of human-machine must be clearly defined so that the transmission of information between different parties are always clearly acted upon without any interference or ambiguity.

With all these constraints in mind, it is important that human participates activiely in the whole system development from developmental till implementation stage so that the information presented and integrated in the human-machine interface are timely and relevant.

The Application of Automation

Over time, the application of automation has taken a more "active" role in the work processes of controllers; moving from collating of various information to integrating this wealth of information "intelligently" and providing assistance to controlling problem.

However, problems still exist in this automated integrating process due to the overwhelming number of areas where information needs to be tapped as seen from the mismatch in information from the paper flight strip containing aircraft detail and what is displayed on the radar console, thus generating problem of duplication.

Further implications of Automation

With automation in an advisory capacity, critical decisions can only be made with agreement from the controller. Hence this advisory role would be more suited in the planning aspect through the manipulating of constraints defined by the controller; this would enable true human-machine cooperation to be achieved with planning handled by the machine (sequencing of aircraft) and executive decisions (traffic deconfliction) still in the hands of the controller, thus retaining human expertise.

Another role machines could partake would be identifying and solving certain simple problem so as to relieve some controller's workload.

Nonetheless, automation of data could lead to Human Factors problems as controllers might not have a good feel regarding the reliability of information. Also certain routine tasks which aid the basic fundamentals of controlling could be eroded. It is thus imperative that all these problems be looked at and catered for in advance in the developmental stage of the system.

Team Function

Traditional team ethos in assistance of controlling by colleagues or supervisors might be rendered irrelevant what with automated assistance enabling controllers to be self sufficient in their constant interaction with the machine.

This would make assessment of competency level in controllers difficult due to reliance of machines with the leaning towards simulators as a means to gauge the proficiency of controllers.

Standardization

The imposing of standardization across communication and controlling still leaves room for debate due to a de-sync between the degree of standardization that should be accepted as it brings about its own set of problems by discouraging human flexibility and its incompatibility of ingrained verbal practice.

Human-Machine Interface and Human Error

Traditional and existing means of the human-machine interface will be changed with automation as new methods in the conveying of information are utilized. This brings about a whole set of new problems as human error becomes less predictable and hence new kinds of human error need to be identified.

The Selection and Training of Air Traffic Controllers

Selection of Applicants

Due to the demanding nature of air traffic, only the selection of the right candidate capable of becoming a proficient controller will lead to the continued safety and efficiency in the job.

This is done by attracting a large pool of applicants through publicity, through extolling how desirable ATC as a profession can be. Thereafter, this large pool can be downsized through impartial yet stringent selection procedure based on Human Factors principle.

Tests

Prior to the administrating of standardized test, particular desirable attributes in controllers are first identified so that they are assessed accordingly in candidates.

However, no one test has gained universal acceptance which would enable it to be totally dependent in controller selection and these tests are generally used as a means to find out more about a particular attribute of the candidate.

Tests are also slowly becoming more automated but care must be taken to ensure candidates receive the proper practice and familiarization with the automated testing procedures unless it would impact on their performance.

These selection process must continuously evolve as and when changes occur in ATC and when new examinable human dimension are discovered so that the process is not made irrelevantly.

Items that could potentially impede controller's performance also need to be looked at and this includes medical histories, addictions, physical handicaps, emotional problem and mental limitation.

To make the selection process holistic, an impartial interview should be conducted to gauge two essential ingredients of ATC success; expression of speech and how good a team player the candidate is.

Training

A proficient controller needs to be cognizant in the fundamental of how ATC is conducted in line with acceptable standards whilst utilizing the required information, methods of communication and equipment. They must also know how to work in a team, detect system failures and how to gauge aircraft performance. All these can only be achieved through proper training.

Pertinent issues which should be addressed by specific Human Factors training include: learning and comprehending procedures related with safe conduct of ATC, how to liaise with controllers and pilots, identifying and preventing human error, knowledge of one's weakness and the motivation to eradicate them through remedial training, garnering of professional attitudes and aptitudes whilst seeking high standards.

The training process must thus allow trainees to make the best use of human capabilities whilst negotiating its limitations in relation to the learning, understanding and remembering processing of information. Only then would it lead to controllers being able to discharge their duties in accordance to highly acclaimed international standards.

For training to be beneficial, it should be conducted in a progressive manner so that trainees are able to master the fundamental principle so that the knowledge they acquire will hold them in good steady when dealing with more complex problem later. How far the trainee has progressed can be tested through assessment along the way.

It is also important to ensure that any ambiguities arising from new forms of material must be properly taught and learned otherwise it would lead to new human error.

Although more self-training packages for students to practice their skills on the computer are in the pipeline, the current flavor in teaching method is through active participation via practical session and simulation. Even on the job training entails honing whatever they have learnt with qualified controllers in real life scenarios at ATC center.

On the job training is vital in the training process as a gauge to check how students deal with the pressure of controlling. Hence it is important that the instructor possesses the necessary experience and acumen in coping with the twin responsibility of controlling real life traffic and imparting ATC skills to the student.

With the advent of more automated and sophisticated systems, it is essential that controllers retain the fundamental ATC practices and procedures through refresher and competency checks so that they are not over-reliant on automation and can still deal with contingencies in the event of system failure.

Due to the fact that currently ATC cannot function autonomously, essential objectives of training should include and identify everything that the controller needs to know, do and say at the appropriate time in the appropriate manner.

Training should hence follow recommended Human Factors procedures and practices that would cater to their individual need and provide an insight into their limitation and capability whilst being able to select the appropriate tools to enhance their work performance. It is also important in instilling good habits, knowledge and skill so that the controller can plan, deconflict, multi-task and deal with contingencies; these skills must be maintained actively and reinforced from time to time. Training must encourage constant scanning and alert techniques so that the controller does not succumb to tunnel vision and become overly preoccupied with certain tasks and neglect other important tasks and must prepare the controller for both types of light and heavy traffic so that he is capable of dealing with each of them confidently and safely.

A sign of good training is whereby controllers are able to deal with all kinds of scenarios through the breeding of self confidence based on past and achieved performance.

When new changes arise in ATC system, it is imperative that there is a careful redefinition of all the consequent changes in the controller's knowledge and procedures. Retraining should be done before the controller encounters these changes in real life scenarios.

However, initial training of new controllers and retraining of qualified controllers may not be the same as retraining involves the unlearning and discarding of familiar knowledge and inappropriate practice ingrained.

The Human Element-Specific Attributes

Recognition of their significance

For Human Factors in ATC to be holistically addressed, besides those that are addressed through the SHEL model, two other categories associated with specific human attributes have to be considered and will be elaborated further.

The first category concerns the effects of ATC on those who work within it; this can be managed by modifying the ATC system. The second category refers to innate human attributes which are independent of ATC environment and to which ATC must therefore be adapted to. In both cases, solutions of particular problem may differ due to the myriad of different causes which potentially might lead to either changes to the system or to the human.

The first category includes attribute such as:

- Stress—Workload of controllers are at the very high end of the stress spectrum due to the demanding nature of the job. Factors that have contributed to stress include the hostile environment involving fear of blame, poor relationship with management, shift work which affect the circadian cycle and modern lifestyle. Although prevention is better than cure, what is needed is correct diagnosis of how stress came about. Some ways to alleviate the problem include redesigning task and reallocating responsibilities by transferring the controller to a less demanding job or to adjust the number of hours of work through proper shift rostering.

- Boredom—Although still not very well comprehended, it is caused by little activity and routinization. Hence there is a need to alleviate this by allowing controllers leeway to plan their workload, providing ample skilled work to do and have groups of people around work to interact while working.

- Confidence and Complacency—Confidence is essential in controlling but is detrimental if it leads to complacency. This can be reduced by constantly dishing out high work level and providing stimulating problem to solve.

- Error Prevention—The error is human, and from this basis although effort have been made to prevent human error, it may not be sensible to predicate the safety of the ATC system on the assumption that every human error can be prevented; it is more important to make the system error tolerant through system design. Although various classification of human error in ATC have been compiled through reported ATC incidents containing details of human errors, an alternative approach worth considering is through procedures which are formulated to remove error and prevent further consequence by classifying error classification based on error task, execution and knowledge.

- Fatigue—Fatigue is an important aspect as it would lead to impairment of judgment and hence safety to be compromised in ATC. Hence, adequate rest break must be provided along with provision of meal break and to be mindful of stretching long, continuous working hours vis-a-vis traffic intensity.

The second category includes:

- Needs at work—Humans, unlike machines have job expectation and it is important to balance their satisfaction, with their aspiration managed for fear of being made redundant as automation is slowly introduced.

- Attitudes—Favorable attitudes towards the ATC environment and as a profession both individually and from outsiders are important for controllers so management must continuously strive to achieve this.

- Functions of Teams—With the introduction of automation, there is a need to look into all facets of team function to restore the optimum match between human and machine.

- Individual differences – The current selection and training processes involves selecting and training candidates of similar characteristics so management know what to expect in their performance while on duty. With automation, this trend may move to honing individual difference and strength so as to tap on their varied potential and background.

A General Human Factors View

The genetic makeup of human's mental faculties in their thought process of understanding and decision making must be factored into the ATC system in order for task competency in training and job-scope to be maximized.

For optimum performance in ATC, there is a need for review at appropriate time so that there is not any degradation in working hours, shift cycles and ergonomic factors at work.

The future is pointing towards greater air traffic which would entail current ATC system which are working at close to or at maximum capacity to be further enhanced and developed.

With this prospect of further development and technological changes, it is necessary to identify and resolve quickly the associated Human Factor consequence.

II Words and Expressions

synopsis	*n.* 提要,概要,梗概
circular	*adj.* 圆形的,环形的
digest	*n.* 文摘;摘要
fraternity	*n.* 同业者;志趣相投者
whilst	*conj.* 当……时;在……的过程中;虽然,尽管
imperative	*adj.* 紧急的,极重要的
interlink	*vt.* (使)连接

entity	n. 实体
prominence	n. 重要;杰出;著名
specify	vt. 具体指明;明确说明;详述
mandatory	adj. 依法必须做的,强制性的,义务的
exponential	n. 指数式增长/增加
collision	n. 碰撞;相撞事故
quantitative	adj. 数量的;与数量有关的;定量的
ergonomics	n. 工效学,人机工程学
spacious	adj. 宽敞的
unimpeded	adj. 无障碍的,无阻挡的
seamless	adj. 流畅的,浑然一体的
console	n. 调节台,操纵台
Anthropometry	n. 人体测量学
alphanumeric	adj. 字母数字混合的;文数式的
decipher	vt. 辨认;解释
articulate	vt. 清楚地表达
transmission	n. 传输,发送,传送
retrain	vt. 接受再培训;对……再训练
myriad	adj. 无数的,极大量的
concise	adj. 简洁的,简练的,简明的
anomaly	n. 异常现象,反常现象
conversant	adj. 熟悉的;有经验的
delve	vt. 探索,探究,查考
attribute	vt. 把……归因于……
plug	n. 插头;塞子,栓
erode	vt. 侵蚀,腐蚀;风化;逐步损害,渐渐削弱
cognizant	adj. 知道的,认识到的,了解的
collate	vt. 核对,校勘
overwhelming	adj. 难以抗拒的,令人不知所措的
strip	vt. 剥去,除去
console	vt. 安慰,慰藉,安抚
duplication	n. 重复

ethos	*n.*	精神特质,道德意识
render	*vt.*	使成为;使变得;使处于
gauge	*n.*	测量仪器,计量器,量规
ingrain	*vt.*	把……深深地印在头脑中;渗入
extol	*vt.*	赞美,颂扬
impartial	*adj.*	不偏不倚的,公正的
stringent	*adj.*	严格的,严厉的
familiarization	*n.*	熟悉,精通;亲密
impede	*vt.*	阻止,妨碍
handicap	*n.*	残障,残疾
liaise	*vi.*	联络,联系
eradicate	*vt.*	根除;消灭
garner	*vt.*	得到,收集
ambiguity	*n.*	模棱两可;不明确
pipeline	*n.*	管道,管线
acumen	*n.*	敏锐,聪明,机智
sophisticate	*adj.*	复杂的;精致的;富有经验的
	vt.	使迷惑;篡改
competence	*n.*	能力,胜任
discard	*vi.*	丢弃
routinization	*n.*	程序化;常规化;惯例化
leeway	*n.*	余地;风压差;偏航;落后
complacency	*n.*	自满;满足;自鸣得意
detrimental	*adj.*	有害的,不利的
taken into account		考虑在内
in spite of		不管,虽然,尽管,不顾
be dealt with		必须处理
self sufficient		自给自足的
as a means to		作为……的一种方法
in the pipeline		在运输中,在进行中,在酝酿中,在准备中

Ⅲ   Exercises

1   Translate the following phrases into English

(1)人机系统

(2)飞行情报区

(3)空管责任

(4)管制员负荷

(5)语言表达

(6)新管制员初始培训

(7)管制员复训

(8)定性和定量的信息

(9)倒班

(10)个体差异

2   Translate the following sentences into Chinese

(1)Future ATC system will attempt to help controllers through more concise and processed information, alerting controllers to anomalies and conflictions and help in predicated future problems by offering solution of projected confliction in flight path.

(2) Prior to the administrating of standardized test, particular desirable attribute in controllers are first identified so that they are assessed accordingly in candidates.

(3)These selection process must continuously evolve as and when changes occur in ATC and when new examinable human dimensions are discovered so that the process is not made irrelevant.

(4)Training should hence follow recommended Human Factors procedures and practices that would cater to their individual need and provide an insight into their limitation and capability whilst being able to select the appropriate tools to enhance their work performance.

(5) Training must encourage constant scanning and alert techniques so that the controller does not succumb to tunnel vision and become overly preoccupied with certain task and neglect other important task and must prepare the controller for both types of light and heavy traffic so that he is capable of dealing with each of them confidently and safely.

IV   Supplement Reading

Human Factors

Introduction

Human factors issues, specifically human errors, contribute to more aircraft incidents and accidents than any other single factor. Human errors include errors by the flight crew, maintenance personnel, air traffic controllers, and others who have a direct impact on flight safety.

What lies behind human error is very frequently inaccurate situational awareness: the failure (for whatever reason) to evaluate an operational or maintenance situation properly. Thus whenever the term human error appears, the reader should keep in mind that situational awareness, or the lack thereof, is usually the dominant factor. This can be a critical problem. As noted in Chapter 2, lack of situational awareness is a key factor in CFIT (controlled flight into terrain) accidents, which are responsible for more fatalities than any other type of aircraft accident.

This chapter discusses the relationship between human factors, environmental factors, and equipment factors in accident and incident; reviews current initiatives to reduce accident and incident associated with human errors or misunderstanding; and recommends steps the FAA can take to improve the effectiveness of its human factors work.

Relationships Between Human Factors, Environmental Factors, And Equipment Factors In Accident And Incident

Human factors are significant contributors in approximately 70 percent of all accident and incident. In a review of several databases, the committee found values in the range of 60 percent to 85 percent. These differences do not reflect on the integrity of the databases; they reflect the databases' different purposes and the understandable difficulties that arise from the substantial overlap of environmental, equipment, and human factors issues. This overlap, which is illustrated in Figure 1, is intrinsic to a complex system with a large number of possible accident and incident sources (primary and contributory). For example, adverse weather (or the threat of adverse weather) can contribute to an accident in many different ways. Weather information is generally but not always accurate; weather information provided to flight crews at dispatch and in flight is generally but not always timely; flight crew decision based on available information are generally but not always made in accordance with prescribed procedures. There is no clear way, and indeed no practical need, to separate entirely environmental from operational factors.

Inaccurate situational awareness by the flight crew can arise in several different ways. Some examples are listed below:

- The flight crew may not have critical data necessary to adequately define its situation, which may lead to inappropriate decision and, ultimately, an accident.
- The flight crew may have the data it needs but misinterpret the data.
- The flight crew may have the data it needs, properly interpret the data, and accurately define the situation, but it may not have the training, skills, or procedures to make proper decision or to carry them out in the time available.

Automated features of flight control system can improve situational awareness by reducing crew workload. However, automated action that compensates for unusual flight condition or equipment malfunction can reduce situational awareness if the automated system masks the presence of abnormalities or does not clearly indicate what action is taking in response.

Aircraft must be designed so that, for all situations the flight crew can reasonably be expected to encounter, it will have the data it needs in an easily recognizable form that facilitates proper decision making. Furthermore, the aircraft should be designed to help the flight crew carry out necessary task, especially in emergencies when things are not as expected and safety depends on quick and correct action by the flight crew. Except for the time pressure typically associated with in-flight emergencies, the same consideration is applied to the actions of maintenance personnel.

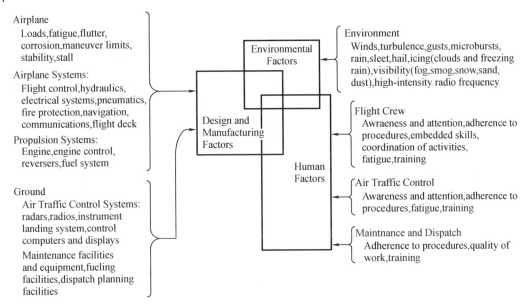

Figure 1   Elements for consideration in safety evaluation

Current Initiatives To Reduce Accident And Incident Associated With Human Error

Many aspects of human factors are associated with the operational safety of commercial airplanes, including the following:

- design factors associated with aircraft control, aircraft system control, warning system, air traffic control system, flight deck, passenger seating and egress, etc.

- operational factors associated with the selection and training of flight crews, crew assignment policies related to the distribution of experienced personnel and the minimization of flight crew fatigue, checks on crew members' health, and policies on preflight information

- maintenance factor related to training maintenance workers; the clarity of maintenance procedure; and designing aircraft equipment and maintenance tool to make it easier for workers to perform maintenance, avoid error, and detect abnormal condition

- national and international regulatory factor associated with airworthiness standard, separation standard, and communications standard

Current processes, which are both thorough and complex, have resulted from a large accumulation of flight experience, analytical and computer studies, and reviews of human factor. All of this information represents a complicated web of interrelated factor that makes it difficult to define a clear and simple road map for progress. Complexity, however, is inherent in many human factor issues.

Figure 2 provides a greatly simplified view of human factor initiatives related to aviation. A much more detailed picture of the breadth and depth of current work and what needs to be done is available in the following publication:

- The detailed report of the FAA Human Factors Team, *Interfaces Between Flight Crews and Modern Flight Deck Systems* (1996), includes more than 50 well formulated recommendations.

- The *Proceedings of the FAA Workshop on Flight Crew Accident and Incident Human Factors* (1995) explores three human factor objectives.

- The April 1997 International Symposium on Aviation Psychology includes more than 300 papers on human factor associated with flight safety.

- The National Aeronautics and Space Administration report, *Principles and Guidelines for Duty and Rest Scheduling in Commercial Aviation* (1996), defines numerous general principles, specific principles, guidelines, and strategies for improving duty and rest scheduling practice.

- The NTSB review, *Flight-Crew-Involved Major Accidents of U. S. Air Carriers*, 1978 *through* 1990 (1992), includes five broad recommendations.

- *Human Factors Guide for Aviation Maintenance* (FAA, 1997), published by the FAA's Office of Aviation Medicine, presents basic concept on reducing human error in maintenance.

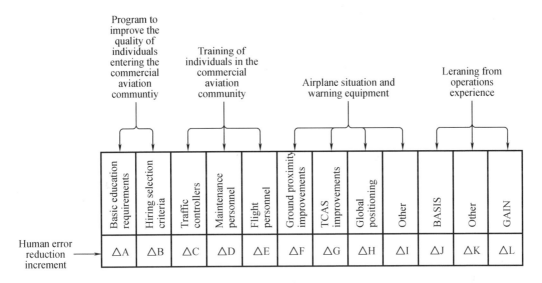

The sum $\triangle A+\triangle B+\triangle C+...+\triangle L$ represents the total contribution available from a reduction in human error.

Figure 2   Current initiatives to reduce human error contributions to accidents/incidents

- A video tape, "Every Day: A Programme about Error Management in Aircraft Maintenance," developed and published by the International Federation of Airworthiness (1997), features Professor James Reason and reviews human factor issues related to maintenance.

Additional work in fields such as cognitive science and fundamental neuroscience is progressing rapidly and is likely to offer valuable insight in the near future. The potential benefit of relying more on cognitive science are explained by James Reason (1990) and, in a more philosophical sense, by David Chalmers (1996). Turning to cognitive science to improve the understanding of issues associated with situational awareness has two major advantages. First, it should encourage the development of processes and systems that would improve the selection and presentation of necessary information, assigning to automated systems of the task that systems do best and allowing people to continue doing the task which people do best. Second, it should help define the type of automation that can reduce the workload of flight crews and air traffic controllers in the crucial moment when a situation must be assessed quickly and accurately.

Improving The Effectiveness Of The FAA's Human Factor Projects

Harnessing the growing body of human factor knowledge will enhance the FAA's effort to reduce the number of incidents and accidents by reducing human error and improving the ability of flight crews and other personnel to prevent accidents associated with other causes. That is one of the tasks of the FAA's Human Factors Study Group. This group appears to be reasonably well coordinated with the JAA(Joint Aviation Authorities)Human Factors Study Group and will operate indefinitely. Close coordination between these two groups is important in an environment that is becoming increasingly aware of the value of international harmonization of airworthiness standards and procedures. Coordinating the work of both groups with similar study groups sponsored by ICAO and other certifying authorities would also be worthwhile.

The membership of the FAA Human Factors Study Group should be reviewed and adjusted, if necessary, to ensure that it has strong representation from the fields of cognitive science and basic neuroscience. Strong representation in these areas would help the study group form a cohesive framework for understanding the very large number of human factor studies that are now under way, and it would enhance the ability of the group to recommend action based on the results of these studies. Some of these studies are associated with enhanced ground proximity warning system, improved traffic collision avoidance system, and other aspect of developing crew-centered cockpit design.

Finding 1. Maintaining situational awareness is the key to prevent the vast majority of serious incidents and accidents associated with human error.

Major Recommendation 4. The FAA should support and accelerate effort (1) to define the minimum data required by the flight crew to maintain adequate situational awareness during all phases of flight and reasonable emergency scenarios (2) to determine how this data can be presented most effectively.

Recommendation 1. The FAA should ensure that its human factor projects, especially the FAA Human Factors Study Group, include strong representation in the fields of cognitive science and basic neuroscience.

Recommendation 2. Advance in understanding human factor should be quickly applied to the key task of reducing the role of human error in incidents and accidents, particularly with regard to improving the situational awareness of operational personnel and the effectiveness of maintenance personnel. The FAA should strongly support its Human Factors Study Group and other projects that contribute to this task.

Unit 14

Risk Management

I Reading Text

PAVE: A Personal Minimum Checklist for Risk Management

By Sarina Houston

We love acronyms, and the pave acronym paves the way for a personal minimum checklist for pilots. Each letter of the acronym stands for a different risk factor associated with flying: personal, aircraft, environment and external pressure. As part of the risk management process, these risk factors should be identified, and the pilot should decide what his or her personal minimum for flight should be based on his self-assessment.

The pave checklist is meant to be used during the preflight planning stage of a flight.

How does aviation define risk management?

Pilots are taught risk management in all aspects of aviation. In recent years, the idea of single-pilot resource management (SRM) has become an important topic in aviation due to the rising number of technologically advanced aircraft (TAA) being manufactured.

Airliners and large aircraft have always had complex systems that required a high level of risk management and safety awareness; now smaller aircraft cockpits are being equipped with these advanced systems, too. In the meantime, the FAA's flight training standards haven't changed much over the years, meaning the same pilot that learned how to fly in an airplane with traditional instrument isn't required to get additional training for an aircraft with advanced equipment on board.

Many aircraft accidents have been the result of not identifying the risk associated with that specific flight before they occur.

Pilots are often caught off-guard by things they should have planned for, such as weather conditions or problems interpreting a piece of advanced technology. The risks associated with flying come from many different places including the pilot himself, the aircraft, the surrounding environment and external pressures involved with each person's unique situation.

These risk factors can affect the safety of flight, and each one should be assessed by the pilot before a flight.

One of the benefit of using the pave checklist is that it allows for pilots to set their personal minimum and stick to them. Each person will have different minimums based on their specific flight experience, health habit and tolerance for stress, to name a few. A pilot's minimum will change over time—as they become comfortable in a particular airplane or environment, for example—but should never be modified or reduced just to rationalize the desire to get off the ground.

Personal

Personal minimum will include pilot health and experience, and can be evaluated in depth with the I'M SAFE checklist. How many hours of sleep do you usually need to function well? Are you healthy? Have you battled any illness or are you on any medication? How much flight experience do you have in the aircraft you're about to fly? How many hours have you flown in the past week/month/year? Are you rusty or stressed? All of these factors can affect your flight.

Aircraft

Is the aircraft airworthy? Did it undergo any inspection recently? Do you have the fuel necessary? Are you comfortable with the weight, balance and performance for the flight?

Do you know the aircraft limitation? Do you have current chart? Is the GPS up-to-date?

Environment

What's the weather like? Are you comfortable and experienced enough to fly in the forecast weather condition? Have you considered all your options and left yourself an "out"? Are you instrument-current? Are you comfortable with the type of approaches available to you? Did you check PIREPs and NOTAMs? Are you comfortable with flying in busy airspace or on edge about the air traffic control situation? Does the aircraft have heat or air conditioning? Are you familiar with the terrain?

External Pressure

Are you stressed or anxious? Is this a flight that will cause you to be stressed or anxious? Is there pressure to get to your destination quickly? Do you have a plan B? Are you dealing with difficult passengers or an unhealthy safety culture?

Are you being honest with yourself and others about your pilot abilities and limitations?

As pilots, we love our checklists. So it's no surprise that there is a self-assessment checklist to assist pilots in determining their own physical and mental health before a flight.

The I'M SAFE Checklist is taught early in flight training and is used throughout a pilot's professional career to assess their overall readiness for flight when it comes to illness, medication, stress, alcohol, fatigue, and emotion.

- I—Illness

The FAA requires most pilots to possess a valid medical certificate for flight, but the occasional medical exam every five years doesn't cover illness such as cold and flu. In the interest of safety, the FAA does regulate this topic loosely by stating that if a pilot has or develops a known medical condition that would prevent him from obtaining a medical certificate, he is prohibited from flying as a required crew member (FAR 61.53).

In addition, FAR 91.3 states that the pilot in command is directly responsible for the operation of the flight. The pilot alone is responsible for ensuring his health is up to par before taking the control.

Colds, allergies, and other common illnesses can cause problem for pilots. From sinus pressure to general malaise, pilots can easily become more of a risk to the flight than an asset.

Before flying, pilots should think about recent or current illness that might affect flight. After the coughing and sneezing subside, a pilot might feel well enough to fly but could still have trouble performing the Valsalva maneuver, which equalizes the pressure inside his ears.

- M—Medication

With illness, it's mostly clear when a pilot should or shouldn't fly. But as illness comes medication, all medications should be scrutinized by both the pilot and his or her doctor before taking it. Many prescription and over-the-counter medications can be dangerous for a pilot to take before flying.

If medication is necessary, pilots should discuss the specific effect of the medication with an aviation medical examiner to determine if it causes mental or physical impairment that would interfere with the safety of flight. Then pilots need to be aware of residual effect of both short-term and long-term use of medications. Even after the medication has been stopped, the effect of it may remain in the body for some time.

So how long should you wait after taking medication to fly? Well, that depends on the drug itself, but the FAA recommends waiting until at least five dosage periods have passed. If the medication is taken once a day, for example, you would wait five days before flying again.

- S—Stress

There are at least three kinds of stress that pilots should be aware of: physiological, environmental and psychological stress.

Physiological is stress in the physical sense. It comes from fatigue, strenuous exercise, being out of shape or changing time zone, to name a few. Unhealthy eating habit, illness, and other physical ailment are included in this category, too.

Environmental stress comes from the immediate surroundings and includes things such as being too hot or too cold, inadequate oxygen level or loud noise.

Psychological stress can be more difficult to identify. This category of stress includes anxiety, social and emotional factors and mental fatigue. Psychological stress can occur for many reasons such as divorce, family problem, financial trouble or just a change in schedule.

A small level of stress can be a good thing, as it keeps pilots aware and on their toes. However, stress can accumulate and affect performance. Also everyone handles stress differently. A source of anxiety for one person might be a fun challenge for another person. It's important for pilots to be able to recognize and evaluate their stressors so they can mitigate risk.

- A—Alcohol

There's no doubt that alcohol and flying cannot be mixed. Alcohol abuse affects the brain, eyes, ears, motor skills and judgment, all of which are necessary component to safe flight. Alcohol makes people dizzy and sleepy which decrease reaction time.

The rules surrounding the use of alcohol while flying are clear: FAR 91.17 prohibits the use of alcohol within the 8 hours before flying, while under the influence of alcohol, or with a blood alcohol content of 0.04% or greater. The FAA recommends that pilots wait at least 24 hours after drinking to get behind the controls.

A pilot should remember, though, that they can follow the "8 hours from bottle to throttle" rule and still *not* be fit to fly. Hangovers are dangerous in the cockpit, too, with effect similar to being drunk or ill: Nausea, vomiting, extreme fatigue, problems focusing, dizziness, etc.

- F—Fatigue

Pilot fatigue is a difficult problem to address completely, as fatigue affects everyone differently. Some people can function well with little sleep; others don't perform well at all without at least ten hours of sleep per night. There's no medical way to address the sleep issue with pilots—each pilot must be responsible for knowing his or her limitations.

The effects of fatigue are cumulative, meaning that small sleep deprivation over time can be dangerous for pilots. Pilots should also take time changes into account, jet lag and day/night scheduling options when managing fatigue.

Although there are FAA regulations and company policies for commercial pilots to help manage fatigue, the responsibility for safety lies with the pilot alone.

- E—Emotion

For some people, emotions can get in the way of behaving in a safe, productive manner. Pilots should ask themselves if they are in an emotionally stable state of mind before departing. Emotions can be subdued and managed most of the time, but they can also resurface easily, especially when faced with a stressful situation.

Most of the time, this type of self-assessment is hard, but pilots need to try to maintain an objective view of themselves to assess their behavior and emotions in a safe way. For example, if a pilot notices that he is unusually angry or impatient while preparing for a flight, he may want to reconsider flying.

Ⅱ Words and Expressions

cockpit	*n.*	驾驶舱
modify	*vt.*	修改,更改,改进,改造;修饰
rationalize	*vt.*	找出辩解理由;进行合理化改革
inspection	*n.*	视察;检查;仔细检查
allergy	*n.*	过敏性反应,过敏症
sneeze	*vi.*	打喷嚏
maneuver	*n.*	熟练的动作;策略,巧计,花招;军事演习
scrutinize	*vt.*	仔细检查,认真查看,详审
prescription	*n.*	处方,药方
dosage	*n.*	剂量,服用量
physiological	*adj.*	生理学的;生理的;生理作用的;生理机能的[亦作 physiologic]
strenuous	*adj.*	艰苦的;费力的,费劲的
ailment	*n.*	小病
inadequate	*adj.*	不够好的,不足的,不强的;不胜任的,不够好的
mitigate	*vt.*	减轻,缓解,缓和
hangover	*n.*	宿醉
dizzy	*adj.*	头晕目眩的;粗心大意的,糊里糊涂的,健忘的
fatigue	*n.*	疲劳,疲乏;杂役
deprivation	*n.*	缺少,匮乏
lag	*vt.*	落后于,拖后;加上保温层
subdue	*vt.*	镇压,制服;克制,抑制
catch/take sb off-guard		措手不及,猝不及防
stick to		紧跟
to name a few		举几个来说,仅举几例
over-the-counter		非处方的
interfere with		妨碍,阻碍,阻挠;干扰

get behind	落在后面,没及时…… 赞同或支持
take into account	把……考虑进去,考虑到;体谅;顾及;重视

Ⅲ Exercises

1 Translate the following phrases into English

(1)与飞行相关的风险因素

(2)风险管理和安全意识

(3)医疗执照

(4)飞行训练标准

(5)规章与公司规定

(6)一个情感稳定的状态意识

(7)降低反应时间

(8)精神和物理损伤

(9)经济困难

(10)氧气缺乏等级

2 Translate the following sentences into Chinese

(1)It comes from fatigue, strenuous exercise, being out of shape or changing time zones, to name a few. Unhealthy eating habit, illness, and other physical ailments are included in this category, too.

(2)A pilot should remember, though, that they can follow the "8 hours from bottle to throttle" rule and still not be fit to fly. Hangovers are dangerous in the cockpit, too, with effect similar to being drunk or ill: Nausea, vomiting, extreme fatigue, problems focusing, dizziness, etc.

(3) Emotions can be subdued and managed most of the time, but they can also resurface easily, especially when faced with a stressful situation.

(4) The effects of fatigue are cumulative, meaning that small sleep deprivation over time can be dangerous for pilots.

(5) Pilots should also take time changes into account, jet lag and day/night scheduling options when managing fatigue.

IV   Supplement Reading

Single-Pilot Resource Management (SRM)

By Sarina Houston

Single-pilot resource management, or SRM, is a derivative of crew resource management (CRM) and is a relatively new term that applies to single-pilot operations. CRM was implemented to help crew members communicate effectively while using all available resources to identify and manage risks before, during and after a flight. Single-pilot resource management is the same thing, but for pilots who operate without fellow crew members.

SRM was implemented as part of the FAA FITS program.

Single-pilot operations are inherently more dangerous than operations that involve crew members. A single person can be more easily overwhelmed when faced with multiple decisions to make. Task management can quickly become difficult for even seasoned pilots when things go wrong. For example, in the same emergency, a dual-pilot crew can divide the responsibilities and tasks in half, and each accomplishes their given task. Airline pilots can be assisted by flight attendants, off-duty crew members and even passengers in emergency situation.

SRM Concept

A single pilot has nobody to help him. The good news is that through SRM, a single pilot is taught to manage workload, mitigate risk, correct errors, and make good decisions—just the same as a crew would do with CRM concept.

- Aeronautical Decision Making (ADM) and Risk Management (RM) SRM training teaches pilots appropriate decision-making strategies and risk management techniques. Each flight has some level of risk to it; pilots should know how to do a risk assessment, how to reduce risk, and how to make decisions based on all available information.

- Task Management (TM) Task management is all about prioritizing and identifying tasks that can be completed before, during and after a flight to ensure efficient operation without task overload.

- Automation Management (AM) Today's flying environment is filled with TAA and glass cockpit, so automation management has become a very important concept. Pilots should practice good AM by programming information into avionics before a flight if possible, and by knowing precisely how their systems operate. An extensive knowledge of automation is extremely important for single pilot.

- CFIT Awareness Controlled flight into terrain (CFIT) continues to be a problem, and a single pilot must identify the risks associated with each flight before, during and after flying. Knowing terrain and aircraft capabilities is essential.

- Situation Awareness (SA) Situational awareness is a no-brainer for single pilots. Pilots must stay aware of their position at all times. It's easy to get confused, especially in the cloud, and lack of situational awareness quickly leads to very bad days. Pilots should use the above concept to help them remain aware of their location, route, altitude, etc. at all times.

The 5 P's

A helpful way for a pilot to assess his or her situation as a single pilot is to utilize the concept of the 5 P's, which is a practical way for the pilot to analyze the risk associated with the element of a flight.

- Plan The pilot should accomplish all preflight planning, and be prepared to adjust the flight plan as necessary during the flight. The plan also involves circumstance surrounding the flight planning process, like gathering weather information and assessing the route.

- Plane   The airplane is obviously a significant element of the flight, and the pilot should assess the risk associated with inoperative equipment and the general shape of the airplane.

- Pilot   The pilot should assess himself with a risk assessment checklist and I'm safe checklist, but should also assess his currency and proficiency, as well as the condition of the flight in relation to his abilities and his personal minimum.

- Passengers   Passengers can present challenges like illness, fear, discomfort, and distraction. It's best for a pilot to plan for passenger challenges ahead of time, like providing them each with water and sick sacks, and briefing them about what will occur.

- Programming   Advanced avionics must be understood completely and programmed correctly.

By assessing each of these items and the variables involved, a pilot can discover and mitigate more risks, and make knowledgeable decision on the spot.

Unit 15

Airline Pilots

I Reading Text

Airline Pilots Talk About Their Personal Relationship

For pilots, marriage can be complicated. Most airline pilots wouldn't trade their job for anything. After all, it beats sitting behind a desk, and it comes with a fantastic view along with many other benefits. But there are also challenges, it can be difficult for a pilot's family and friends to understand what exactly they are up against while they're in training or on trips.

How hard can it be to fly around the world, indulging in drinks at hotel bars with fellow crew members?

And why does their schedule always have to be so complicated? I caught up with a few airline pilots on Facebook, and they sounded off on pilot marriages and relationships.

Sarah E is a first officer for a major airline. She says it's hard for outsiders to understand what pilots go through. "It's hard for people who don't live the airline life to understand it. They think that while we are away that we are on vacation and partying. It's difficult to convey the amount of work we do it's fatiguing and challenging, especially for a wife and mother. Sleeping in a hotel and living out of a bag isn't the most fun, but we are pilots and have a passion for what we do. It's in our blood, and it's part of who we are."

The Job Puts a Strain on Relationship and Marriage

For an unlucky number of pilots, their relationships or marriages end due to one or more of these challenges leading to misunderstanding.

Some of these can be blamed on the rigors of the job, which is difficult for any non-pilot to comprehend. And those that are trying to make a marriage work often spend their time explaining the ins and outs of the pilot career to their significant others left behind, often left alone to deal with the challenges of raising a family.

The details, like why pilots spend an enormous amount of money on fast food and why they were scheduled during the family's yearly vacation to Hawaii even though they bid for a different schedule, can become sources of contention, and often, family members behind feel left out and misunderstood themselves.

"This is likely a major reason for pilot divorce—the lack of comprehension on what the job entails," says Melinda W. , a married first officer at a major airline. "One captain clued me into best management practices for a content spouse", "Remember, the weather is always crappy, the hotel is a dump & the crew is a bunch of idiots. "

"Your spouse doesn't want to hear you're having a good time on a trip while they're home dealing with a backed up toilet, a car problem, a sick kid, shoveling snow, or the dog got sprayed by a skunk!" 、"Maybe so, but spouses don't need to be protected from the daily lives of pilots. "

Pilots just need to convey the truth, that while the weather in Hawaii is nice and yes, they did enjoy a cocktail at the hotel bar, and that they're exhausted and yes, still committed to their marriages. And even though they try to explain it, many pilots wish their significant others understood what happens around them every time they fly.

As a first officer at a major airline, Evelyne T. knows the process well, too. "It's a long-term education process. I found that by talking and telling stories and explaining in detail, openly and candidly my work experiences, trials, tribulations, and adventures I can give my family members and friends a window into my world... I don't ever dumb anything down and give lots of backgrounds. "

For others, their best-laid relationship plans didn't work out in the end. A number of pilots responded to my request for information to say their significant other ran off with a flight attendant, or that it just didn't work out for one reason or another. Life happens. Schedules don't line up. Dreams get in the way. And for airline pilots, it's not difficult to see why.

Training

An airline pilot's hectic schedule begins right away, usually during simulator training.

If the pilot is coming out of the military, the training environment is one that they've probably experienced before. But if they built their experience in the civilian world-flight instructing or towing banners or something similar-they're just as new as the rest of the family and probably a bit clueless about the process themselves.

But one thing is for sure: It's called a "fire hose" for a reason. Days are long, the books are thick, and the absorption rate of the material is fast. It's intense. Pilots are expected to learn a huge amount of material in a very short time with little time between lessons.

They go to class all day, maybe grab dinner with their new coworkers at night, review notes for an hour or two, go to bed and then repeat the process the next day. There is very little time for anything else, leaving family members wondering why their husband or wife all of the sudden checked out.

And it's true-pilots often put their partners on hold as they check out of family life and check into a crappy hotel for a few months. Luckily, training is temporary. And it's worth it when they put on that crisp new uniform and epaulets.

Downtime

Once a pilot is done with same training, they often just want to be decompressed. If their partner tries to hand them a "to-do" list, they'll sigh. If their partner makes them breakfast with the hope that they'll join them, they'll sleep in. And if asked where they want to go for dinner, they might respond with "I don't care." Information overload, constantly being in a leadership position and the decision-making faced on the job leaves pilots in a zombie-like anti-decision-making state of mind. They don't care where you eat. They'll eat anything at this point. . . except maybe McDonald's.

On Reserve

After training, a pilot's time home is often brief, and then they'll be off to their reserve location, which means that they have to live near the airport in case they are called upon to fly. It is also a temporary situation, fulfilled while the pilot waits to "fly the line" at their regular gig, but that doesn't mean it's not challenging. If they're lucky, the reserve base is nearby.

The majority of pilots, however, live in a crash pad in another city while on reserve. While crash pad living might sound like a party, your pilot is just as annoyed as you might be about this situation. He or she is living with a host of other male or female pilots and flight attendants who are loud, up at all hours and also cranky that they're not at home with their families. It's not the most glamorous life.

On the Line

After a few months on reserve, pilots get a spot flying the line, which means they can bid on their schedule and be at home when they aren't flying. Junior pilots-those low on the seniority list-will fly nights and weekends and any other shift that the senior pilots don't want to bid. It is also temporary and is dependent upon how quickly pilots are retiring and how fast new pilots are being hired.

And even line pilots have their challenges. The job can be exhausting, both mentally and physically. Flying, in and of itself, is mentally fatiguing. Pilots are responsible for hundreds of lives in a single flight, and they make important decision about the safety of those flights. Add the trip through the time zones and a bad airport diet, and the body quickly fatigues, too.

And then there are the relationships they're working to uphold back home. Like anyone else, pilots have lives beyond flying, and there is often an added stress of worrying about kids, spouses and finances while they're away. Jealousy can play a role in a pilot's marriage (or lack thereof), too. Maintaining a friendly relationship with crew members of the opposite sex without causing a significant other to become jealous can be a challenge. And then there's commuting.

Commuting

Commuting is a part of the job for many pilots and involves them flying to their assigned domicile before their schedule even begins and is done on their own time, often adding a day to the beginning of the pilot's scheduled trip. And then they have to commute home, adding a day to the end of the trip, too. By the time a pilot gets home, he or she might not want to leave home, which is why a pilot might balk at the idea of taking a family vacation on his or her week off. Often the last thing he or she wants to do is hopping on another airplane.

A pilot that has seniority can bid for a schedule that suits their need, making it possible for them to finally be home on Christmas and attend important school function for their kids. And the good news for all pilots is that when they're home, they're home. Their time is their own when they're off the clock, which is not something that is true for many other professions.

Relief

If a pilot's marriage can last through the random scheduling, the missed holidays, the jealousies and other various challenges involved with becoming an airline pilot, then maybe, just maybe, they'll see relief. As an airline pilot gains seniority, he or she will be able to gain more control over his or her schedule, allowing more scheduled time for family and other hobbies. As seniority comes a pay increase, and any money argument that were there before may subside. And eventually a pilot will be able to be home on holidays and keep important dates.

The pilot lifestyle is challenging. There are many happily married pilots out there, but the secret to a happy marriage is probably less about pilot scheduling than it is about basic marriage virtue. Understanding the pilot's lifestyle is just the beginning.

II Words and Expressions

outsider	*n.* 局外人士;无望获胜者
rigour	*n.* 严酷;严密,缜密,严谨
enormous	*adj.* 巨大的,庞大的
divorce	*n.* 离婚
crappy	*adj.* 差劲的,很糟的
spray	*vt.* 喷,喷洒,飞溅,飞散
trial	*n.* 审判,审理
tribulation	*n.* 苦难,艰辛
dumb	*adj.* 愚蠢的
hectic	*adj.* 兴奋的,狂热的;忙碌的
hose	*n.* 软管,胶管,水龙带;连裤袜;长筒袜
grab	*vt.* 攫取,抓住;赶紧,抓紧

crisp	*n.* 油炸薯片
epaulet	*n.* 肩饰
gig	*n.* 演奏会;滑稽表演;短时的工作
cranky	*adj.* 奇怪的,古怪的;脾气坏的
glamorous	*adj.* 有魅力的,令人向往的,奢华的
reserve	*vt.* 储备;保留;预约
seniority	*n.* 年资,资历;年长;职位高
uphold	*vt.* 支持,维护;维持原判
jealousy	*n.* 妒忌,嫉妒
domicile	*n.* 住处,住所
subside	*vi.* 逐渐减弱;平静下来,平息

Ⅲ Exercises

1 Translate the following phrases into English

(1)副驾
(2)模拟机训练
(3)婚姻伦理
(4)长期的教育过程
(5)精神疲劳
(6)备份
(7)条幅广告
(8)信息过载
(9)结算
(10)领导职位

2 Translate the following sentences into Chinese

(1) Pilots just need to convey the truth, that while the weather in Hawaii is nice and yes, they did enjoy a cocktail at the hotel bar, and that they're exhausted and yes still committed to their marriages.

(2) After training, a pilot's time home is often brief, and then they'll be off to their reserve location, which means that they have to live near the airport in case they are called upon to fly.

(3) Commuting is a part of the job for many pilots and involves them flying to their assigned domicile before their schedule even begins and is done on their own time, often adding a day to the beginning of the pilot's scheduled trip.

(4) By the time a pilot gets home, he or she might not want to leave home, which is why a pilot might balk at the idea of taking a family vacation on his or her week off.

(5) There are many happily married pilots out there, but the secret to a happy marriage is probably less about pilot scheduling than it is about basic marriage virtue. Understanding the pilot's lifestyle is just the beginning.

IV Supplement Reading

How Pilots Use Air Navigation to Fly

By Sarina Houston

Updated November 20, 2017

Air navigation is accomplished by various methods. The method or system that a pilot uses for navigating through today's airspace system will depend on the type of flight that will occur (VFR or IFR), which navigation systems are installed on the aircraft, and which navigation systems are available in a certain area.

Dead Reckoning and Pilotage

At the most simple level, navigation is accomplished through ideas known as dead reckoning and pilotage.

Pilotage is a term that refers to the sole use of visual ground references. The pilot identifies landmarks, such as rivers, towns, airports, and buildings and navigates among them. The trouble with pilotage is that references aren't often easily seen and can't be easily identified in low visibility condition or if the pilot gets off track even slightly. Therefore, the idea of dead reckoning was introduced.

Dead reckoning involves the use of visual checkpoints along with time and distance calculation. The pilot chooses checkpoints that are easily seen from the air and also identified on the map and then calculates the time it will take to fly from one point to the next based on distance, airspeed, and wind calculations. A flight computer aids pilots in computing the time and distance and the pilot typically uses a flight planning log to keep track of the calculation during flight.

Radio Navigation

With aircraft equipped with radio navigation aids (NAVAIDS), pilots can navigate more accurately than with dead reckoning alone. Radio NAVAIDS comes in handy in low visibility condition and acts as a suitable backup method for general aviation pilots that prefer dead reckoning. They is also more precise.

Instead of flying from checkpoint to checkpoint, pilots can fly a straight line to a "fix" or an airport. Specific radio NAVAIDS are also required for IFR operation.

There are different types of radio NAVAIDS used in aviation:

- ADF/NDB: The most elementary form of radio navigation is the ADF/NDB pair. An NDB is a nondirectional radio beacon that is stationed on the ground and emits an electrical signal in all directions. If an aircraft is equipped with an automatic direction finder (ADF), it will display the aircraft's position in relation to the NDB station on the ground. The ADF instrument is basically an arrow pointer placed over a compass card-type display. The arrow always points in the direction of the NDB station, which means that if the pilot points the aircraft in the direction of the arrow in a no-wind situation, he will fly directly to the station.

  The ADF/NDB is an outdated NAVAID, and it's a system prone to errors. Since its range is line-of-sight, a pilot can get erroneous reading while flying in mountainous terrain or too far from the station. The system is also subject to electrical interference and can only accommodate limited aircraft at once. Many are being decommissioned as GPS becomes the primary navigation source.

- VOR: Next to GPS, the VOR system is probably the most commonly used in the world. VOR, short for VHF Omnidirectional Range, is a radio-based NAVAID that operates in the very-high-frequency range. VOR stations are located on the ground and transmit two signals—one continuous 360-degree reference signal and another sweeping directional signal.

  The aircraft instrument (OBI) interprets the phase difference between the two signals and displays the results as a radial on the OBI (omni-bearing indicator) or HSI (horizontal situation indicator), depending on which instrument the aircraft uses. In its most basic form, the OBI or HSI depicts which radial from the station the aircraft is located on and whether the aircraft is flying toward or away from the station.

  VORs are more accurate than NDBs and are less prone to errors, although the reception is still susceptible to line-of-sight only.

- DME: Distance Measuring Equipment is one of the most simple and valuable NAVAIDS to date. It's a basic method using a transponder in the aircraft to determine the time it takes for a signal to travel to and from a DME station. DME transmits on UHF frequencies and computes slant-range distance. The transponder in the aircraft displays the distance in tenths of a nautical mile.

  A single DME station can handle up to 100 aircraft at one time, and they usually co-exist with VOR ground stations.

- ILS: An instrument landing system (ILS) is an instrument approach system used to guide aircraft down to the runway from the approach phase of flight. It uses both horizontal and vertical radio signals emitted from a point along the runway. These signals intercept to give the pilot precise location information in the form of a glideslope—a constant-angle, stabilized descent path all the way down to the approach end of the runway. ILS systems are widely in use today as one of the most accurate approach system available.

GPS

The global positioning system has become the most valuable method of navigation in the modern aviation world. GPS has proven to be tremendously reliable and precise and is probably the most common NAVAID in use today.

The global positioning system uses 24 United states. Department of Defense satellites to provide precise location data, such as aircraft position, track, speed, and to pilots. The GPS system uses triangulation to determine the aircraft's exact position over the earth. To be accurate, a GPS system must have the ability to gather data from at least three satellites for 2-D positioning, and 4 satellites for 3-D positioning.

GPS has become a preferred method of navigating due to the accuracy and ease of use. Though there are errors associated with GPS, they are rare. GPS system can be used anywhere in the world, even in mountainous terrain, and they aren't prone to the errors of radio NAVAIDS, such as line-of-sight and electrical interference.

Practical Use of NAVAIDS

Pilots will fly under visual flight rules (VFR) or instrument flight rules (IFR), depending on the weather condition. During visual meteorological conditions (VMC), a pilot might fly by using pilotage and dead reckoning alone, or he might use radio navigation or GPS navigation techniques. Basic navigation is taught in the early stage of flight training.

In instrument meteorological conditions (IMC) or while flying IFR, a pilot will need to rely on cockpit instrument, such as a VOR or GPS system. Because flying in the cloud and navigating with these instruments can be tricky, a pilot must earn an FAA Instrument Rating to fly in IMC condition legally.

Currently, the FAA is emphasizing new training for general aviation pilots intechnologically advanced aircraft (TAA). TAA are aircraft that has advanced highly technical system on board, such as GPS. Even light sport aircraft are coming out of the factory with advanced equipment these days. It can be confusing and dangerous for a pilot to attempt to use these modern cockpit system in-flight without additional training, and current FAA training standard hasn't kept up with this issue.

The FAA's updated FITS program finally addressed the issue, although the program is still voluntary.

Unit 16

Meteorology

I Reading Text

New Meteorological Services Supporting Air Traffic Management

The international air navigation system is presently undergoing a paradigm shift: one that is moving it away from past Air Traffic Control (ATC) environment to the more integrated and collaborative Air Traffic Management (ATM) systems now needed to meet aviation's need in the 21st century.

These requirements have stemmed from the more or less continuous growth in aviation and the ever present risk that the capacity of Regional air navigation systems may soon be exceeded by operator demand. This issue is presently of particular importance in the European (EUR) and in North American (NAT) Regions, but it has also become an increasingly urgent priority in the Asia/Pacific (APAC) Region, where the number of intra-APAC air travelers has recently surpassed associated domestic passenger total from the North American market, making today's Asia-Pacific the world's largest aviation market.

The goal of the changes now under development is to ensure that ICAO's vision of a safe, secure, efficient and environmentally sustainable air transport system will continue to be available to all aviation stakeholders at the global, Regional and national levels. The implementation of a new ATM system that will make maximum use of the enhanced capabilities provided by advances in science and technology, as well as allowing for the effective sharing of available information on the basis of Collaborative Decision Making (CDM), is a mandatory component on the path to this objective. The Next Generation Air Transportation System (NextGen) and Single European Sky ATM Research (SESAR) initiatives are the corresponding programs now ongoing in the USA and Europe to help effectively address this challenge.

It is under this evolving environment that the concept of Meteorological Services in the Terminal Area (MSTA) was conceived of by the Commission for Aeronautical Meteorology (CAeM) of the World Meteorological Organization (WMO) in its 13th Session in 2006. In recent years, with increasing air traffic leading to issues of airport and route capacity limit, as well as the advancement of meteorological sciences such as numerical weather prediction and "nowcasting" technique, different meteorological product tailored for supporting Air Traffic Management (ATM) have been developed.

These development has occurred in parallel across various Regions in order to address the gap between the data products stipulated in ICAO's existing standards and guidance and the newer and evolving 21st century ATM user needs for meteorological information. Currently, ICAO Annex 3 stipulates meteorological data products such as the Aerodrome Forecast (TAF), Trend-type Landing Forecast (TREND) and Aerodrome Warnings, which are presented to the users in highly-condensed codes in textual format.

The coded aspect of this weather data was a necessity in the mid-20th century, primarily to overcome the severe bandwidth limitation in legacy telecom system. It has become a severe constraint for meteorologists as they seek to convey the specific details of available weather information to modern aircraft operators.

A case in point is how convective weather, which impacts on busy approach areas, flight routes, corner posts and fixes over the wider terminal area, is already causing significant impact to ATM operations and capacity. Currently this cannot be addressed by the regulated products which only provide generic weather information for the aerodrome—i. e. within approximately eight kilometers of its center. Other products under development, in trial, or already in operational use at some airports in the APAC Region, include wind forecast over approaches, crosswind probability forecast for runways, strike probability for tropical cyclones, etc.

In more advanced applications of these weather products, specific information of the weather impact on air traffic capacity is also generated and provided to ATM and airline users. One such example is the Air Traffic Meteorology Center (ATMetC) of Japan.

To address these new and evolving ATM user need and to avoid the costly parallel development of similar weather products of varying and confusing data format, an expert team has been set up by the WMO in order to work closely with ICAO to develop a proper MSTA proposal. This proposal would be based on commonalities in similar products developed thus far and would also recognize current technical capabilities and limitations.

It is envisaged that this new MSTA data product would provide forecast of weather element critical to aviation in the wider terminal area, along with longer lead times and much finer resolution in space (both the horizontal and vertical domains) and time (especially in the first couple of hours of the forecast) compared to currently available aviation MET products. While MSTA is intended primarily for busier airports and terminal areas, it is also envisaged as a significant enhancement to aviation safety in general.

The new MSTA will be produced in a digital, gridded format, initially being available as a web-based color graphic with alerting criteria. It will provide common situational awareness for data sharing by different user groups in support of CDM, and could be supplemented by textual description as appropriate and simplified/condensed to facilitate uplink to aircraft cockpits.

At this early stage in its development, the MSTA product will focus on forecast of convection, winds, low ceiling/visibility and winter weather. In addition, probability attributes of the various weather element will be included as possible input for user decision support system. A number of core experts from various Regions are working together on MSTA development, including several from the APAC Region (Australia, Hong Kong/China and Japan).

Prototypes of convection and wind product have already been developed and were presented to the 14th Session of the CAeM held in Hong Kong, China in February 2010.

To facilitate further development of MSTA prototypes and input from the aviation community, a web site (http://www. msta. weather. gov. hk/) is currently being hosted by the Hong Kong Observatory with access available to WMO members and aviation users. In addition to the WMO expert team, a new Task Force on MSTA User Needs was also set up by the CAeM to strengthen engagement with aviation user communities, focusing on user need and gathering feedback on the MSTA concept.

On the ICAO side, the Aerodrome Meteorological Observation and Forecast Study Group (AMOFSG) has established an ad hoc group to work closely with the WMO Task Force to coordinate input from the requirement perspective. Its work programme also includes consultation with the ICAO Air Traffic Management Requirement and Performance Panel (ATMRPP).

The objective of these efforts is to develop a detailed MSTA proposal, supported by ICAO and the aviation user communities, ready for endorsement by the next Conjoint ICAO MET/AIM Divisional Meeting/WMO CAeM Session (currently scheduled for 2014) and reflected in ICAO Annex 3.

II   Words and Expressions

Meteorological	*adj.* 气象的；气象学的
undergoing	*vt.* 经历，经受；忍受
paradigm	*n.* 范例
stem	*n.* 干；茎；船首
	vt. 阻止；除去……的茎；给……装柄
	vi. 阻止；起源于某事物；逆行
exceed	*vt.* 超过；胜过
	vi. 超过其他
urgent	*adj.* 紧急的；急迫的
sustainable	*adj.* 可以忍受的；足可支撑的；养得起的；可持续的
implementation	*n.* 实现；履行；安装启用
mandatory	*adj.* 强制的；托管的；命令的
	n. 受托者（同 mandatary）
component	*n.* 成分；组件；元件
	adj. 组成的，构成的
initiative	*n.* 主动权；首创精神；新方案；倡议
	adj. 主动的；自发的；起始的
nowcasting	*vt.* 即时预报
stipulate	*v.* 规定；保证
	adj. 规定的
stipulate	*vi.* 规定；保证

	vt. 规定;保证
	adj. 有托叶的
condense	*vi.* 浓缩;凝结
	vt. 使浓缩;使压缩
bandwidth	*n.* 带宽,频带宽度
legacy	*n.* 遗赠,遗产
convective	*adj.* 对流的;传递性的
trial	*n.* 试验;审讯;努力;磨炼
	adj. 试验的;审讯的
crosswind	*n.* 侧风
cyclone	*n.* 旋风;气旋;飓风
capability	*n.* 才能,能力;性能,容量
element	*n.* 元素;要素;原理;成分;自然环境
finer	*adj.* 出色的,好的
envisage	*vt.* 正视,面对;想象
grid	*n.* 网格;格子,栅格;输电网
textual	*adj.* 本文的;按原文的
facilitate	*vt.* 促进;帮助;使容易
uplink	*n.* 上行线路;向上传输
	vt. 向上传输;从地面输送信息
	adj. 上行线路的;向上传输的
ceiling	*n.* 云高
panel	*n.* 仪表板;嵌板
	vt. 嵌镶板
endorsement	*n.* 认可,支持;背书;签注
annex	*n.* 附件
more or less	或多或少
corner post	角柱;井架大腿底柱;对角立柱;角杆

III Exercises

1 Translate the following phrases into English

(1)终端区气象服务

(2)航空气象学委员会

(3)飞行航线

(4)进场区域

(5)空中交通气象中心

(6)机场气象观测与预报研究小组

(7)趋势型着陆预报

(8)机场例行报告

(9)对流天气

(10)低云低能见度天气

2 Translate the following sentences into Chinese

(1) In recent years, with increasing air traffic leading to issues of airport and route capacity limits, as well as the advancement of meteorological sciences such as numerical weather prediction and "nowcasting" techniques, different meteorological products tailored for supporting Air Traffic Management (ATM) have been developed.

(2) The coded aspect of this weather data was a necessity in the mid-20th century, primarily to overcome the severe bandwidth limitation in legacy telecom system.

(3) It has become a severe constraint for meteorologists as they seek to convey the specific details of available weather information to modern aircraft operators.

(4) To address these new and evolving ATM user need and to avoid the costly parallel development of similar weather products of varying and confusing data format, an expert team has been set up by the WMO in order to work closely with ICAO to develop a proper MSTA proposal.

(5) It will provide common situational awareness for data sharing by different user groups in

support of CDM, and could be supplemented by textual description as appropriate and simplified/condensed to facilitate uplink to aircraft cockpits.

IV   Supplement Reading

Climate Change Could Get You Bumped from a Future Flight

Major airports will see more frequent takeoff weight restriction in the coming decades due to increasingly common hot temperature.

*   By Ethan Coffel, Radley Horton

Figure 1   Alessandro Caproni Flickr (CC BY 2.0)

The following essay is reprinted with permission from The Conversation, an online publication covering the latest research.

Hot weather has forced dozens of commercial flights to be canceled at airport in the Southwest this summer. This flight-disrupting heat is a warning sign. Climate change is projected to have far-reaching repercussion—including sea level rise inundating cities and shifting weather pattern causing long-term declines in agricultural yield. And there is evidence that it is beginning to

209

affect the takeoff performance of commercial aircraft, with potential effect on airline cost.

National and global transportation system and the economic activity they support have been optimized for the climate in which it all developed: Machines are designed to operate in common temperature ranges, logistical plans depend on historical weather pattern and coastal land development is based on known flood zones. In the aviation sector, airport and aircraft are designed for the weather condition experienced historically. Because the climate is changing, even fundamental infrastructure element like airports and key economic sectors like air transportation may need to be redesigned and reengineered.

As scientists focused on the impacts of climate change and extreme weather on human society and natural ecosystems around the world, our research has quantified how extreme heat associated with our warming climate may affect flights around the world. We've found that major airports from New York to Dubai to Bangkok will see more frequent takeoff weight restrictions in the coming decades due to increasingly common hot temperature.

Climate Changes Flights

There is robust evidence that extreme events such as heat wave and coastal flooding are happening with greater frequency and intensity than just a few decades ago. And if we fail to reduce greenhouse gas emission significantly in the next few decades, the frequency and intensity of these extremes is projected to increase dramatically.

The effect on aviation may be widespread. Many airports are built near sea level, putting them at risk of more frequent flooding as oceans rise. The frequency and intensity of air turbulence may increase in some regions due to strengthening high-altitude wind. Stronger wind would force airlines and pilots to modify flight lengths and routings, potentially increasing fuel consumption.

The July heat-related Phoenix flight cancellation happened at least in part because airlines' operational manual didn't include information for temperature above 118 degrees Fahrenheit—because that kind of heat is historically uncommon. It's another example of how procedures may need to be updated to adapt to a warmer climate.

Flying In The Heat

High air temperature affects the physics of how aircraft fly, meaning aircraft takeoff performance can be impaired on hot days. The amount of lift that an airplane wing generates is affected by the density of the air. Air density in turn depends mostly on air temperature and elevation; higher temperatures and elevations both reduce density.

Hot air is less dense than cooler air. That affects the amount of lift an airplane can generate. The Conversation (via Piktochart), CC BY-ND.

The lower the air density, the faster an airplane must travel to produce enough lift to take off. It takes more runway to reach a higher speed, and depending on how long the airport's runway is, some airplanes might risk running out of room before reaching sufficient speed. When this occurs, the only immediate option is to reduce the aircraft's weight to lower its required takeoff speed—by removing passengers, luggage and cargo. This is referred to as a weight restriction.

Weight restrictions happen now, especially in hot places like Phoenix and Dubai and at airports with short runways like New York LaGuardia and Washington D. C. Reagan National, but our research suggests that they may become much more common in the future.

Global temperatures have been steadily rising for decades, and they will almost certainly continue to do so. In some regions, there is evidence that the hottest temperature may increase at a faster rate than the average, further stacking the deck in favor of extreme heat. These hotter temperature will reduce air density and make it much more likely weight restriction which is needed for flights taking off during the hottest parts of the day.

The frequency and magnitude of weight restriction is projected to increase—in some locations, the number of days requiring at least some amount of weight restriction for certain aircraft could double or triple, perhaps covering 50 or more days per year.

The Economccs Of Adaptation

On most affected flight, the amount of cargo, passengers and fuel that must be removed to allow for takeoff will usually be small—between 0.5 percent and 4 percent of the total load. That means fewer paying customers on airplanes, and less cargo on board. When those restriction add up across the global air transport system, the cost can be significant.

Carrying just a fraction of a percent fewer passengers or less cargo can add up to millions of dollars in lost revenue for an airline over years of operation. That makes even small weight restrictions a concern in such a highly competitive and optimized industry. These limits could disproportionately affect long-haul flight, which require large fuel load and often take off near their maximum weight.

There are ways that airlines could mitigate increasing weight restriction. The most feasible is to reschedule some flight to cooler hours of the day—although with air traffic increasing and many airports already operating near capacity, this could prove to be difficult.

Another potential solution is to build longer runways. But that's not always possible: Some airports, like New York LaGuardia, are on coastlines or in dense urban environment. Even where a longer runway is technically possible, buying the land and expanding an airport's physical area may be expensive and politically difficult.

Aircraft could be optimized for takeoff performance, but redesigning aircraft is extremely expensive and can take decades. Manufacturers are always working to build planes that are lighter and more fuel-efficient. In the future, those efficiency improvements will be necessary just to maintain today's performance.

Broader Implication

These changes are merely examples of the countless procedure, processes and equipment requirement that will have to be adjusted for a changing climate. Even if those adaptations are successful, they will take effort and money to achieve.

Many sectors of the economy, including the aviation industry, have yet to seriously consider the effects of climate change. The sooner, the better: Both airport construction and aircraft design take decades, and have lasting effect. Today's newest plane may well be flying in 40 or 50 years, and their replacement are being designed now. The earlier climate impacts are understood and appreciated, the more effective and less costly adaptations can be. Those adaptations may even include innovative way to dramatically reduce climate-altering emission across the aviation sector, which would help reduce the problem while also responding to it.

Unit 17

Aircraft Fuel

I Reading Text

Fuel Starvation: Why Do Pilots Run Out of Fuel?

By Sarina Houston

In aviation, like in so many other things, we often scratch our head over the causes of aircraft accidents. Pilots are human, yes, but things like running out of fuel or flying into the side of a mountain just make you wonder what in the world these particular humans are thinking. These types of accidents are common enough that the NTSB has issued special warning about them, even distinguishing them as "special emphasis areas" for pilot training.

In the pilot training world, this means that flight instructors spend extra time on these topics and every check ride with an FAA designated examiner will certainly include at least a discussion on controlled flight into terrain and fuel management.

On November 28th, 2016, an Avro RJ85 carrying a Brazilian Soccer team crashed in Colombia, killing 71 people. Speculation arose in the immediate aftermath that the aircraft crashed due to fuel starvation, and the question piled up. How can two trained airline pilots flying together run out of fuel?

Mechanical problems are rare with all of the redundancy in place, and even in the case of a fuel leak, the pilots should have noticed in time to fly the airplane to a nearby airport. From the last radio transmission made by the crew, it would seem as if they didn't realize how dire their fuel situation was. We may never know what happened exactly to LaMia Flight 2933, but it leaves us with the question: *why do pilots still run out of fuel?*

In flight training, we pay particular attention to these special emphasis areas, and we stress to students that running out of fuel happens too often by the way for anyone to get comfortable with the idea that they'll never be the one to run out of fuel. We're always double checking and triple checking fuel, talking about the reason some pilots run out of fuel, and examining decision making when it comes to fuel management, fuel stops, alternate airports, and fuel reserves.

And then there are checklists. When we preflight the aircraft in flight training, one of the first things a pilot is taught to check is the fuel levels (mostly so that if we need more fuel, we can call the fuel truck early on or plan more time to stop at the self-serve pumps on the way out, but also clearly to ensure we have enough fuel to complete the flight). With two pilots in the airplane, both pilots are taught to check the fuel gauge and then visually check for fuel when possible to ensure that there is, indeed, the appropriate amount of fuel and the amount somewhat agrees with the fuel gauge (Fuel gauge in general aviation aircraft are known to be less than accurate much of the time). In addition to these initial check, there is a preflight checklist that requires the pilot to check the fuel quantity and drain a sample of fuel from the tanks to make sure it's not contaminated. And during a flight, the cruise and descent checklist often call for the fuel to be monitored or the fuel tank to be switched.

Our flight planning process, when done correctly, should include a close look at fuel planning, including the starting fuel amount, the power setting and coinciding fuel burn for each phase of flight.

And by law, we are required to carry more than enough fuel to make it to our destination, as well as an alternate airport when necessary, plus an addition 30 minutes or 45 minute's worth of fuel for day and night flight, respectively.

Finally, in many aircraft, there are, indeed, proper fuel indicator, fuel flow gauge, and even "LOW FUEL" annunciator light on the panel of most aircraft.

So why after all the planning, checklists, system safety, and an emphasis on fuel management, do pilots simply run out of fuel? Well, like all things that seem simple from the outside, it turns out that it's not that simple.

Fuel starvation in airplane happens for a variety of reasons, most of which are just plain human error.

Improper Planning

Improper planning is probably the largest excuse for running out of fuel.

And even after the fact, the pilot rarely acknowledges that his planning was incomplete or just flat-out wrong, because in their mind, they did everything they knew to do to plan, but "luck" was against them. There are many people who run into bad luck, but there are way more people who just don't plan well. Or maybe they don't plan at all. Perhaps they've just always had enough fuel and luck on their side to make them confident that fuel won't just run out, and they've gotten lazy about flight planning in general. Or maybe they plan the fuel properly to get to their destination, but don't plan for an alternate when necessary.

Fuel Mismanagement

Fuel mismanagement occurs when the pilot forgets to switch fuel tank when necessary, or switches to the wrong fuel tank, or just doesn't monitor the fuel burn during a flight. Much of the time, the problem stems from a lack of understanding of the fuel system itself.

Computational Error

Rarely a pilot will make a blatant computational error by moving a decimal one place or just interpreting a fuel chart incorrectly. If the planned fuel burn is 16.8 gallons per hour and the pilot plans his flight using 1.68 gallons per hour instead, he will clearly be burning more fuel than planned. Most of the time, the pilot or another crew member, or even a computer catches the error at some point soon enough to avert disaster, but not always.

Poor Decision-Making

Fuel starvation is often a direct result of poor decision making in multiple areas of the flight. Maybe the pilot didn't get a proper weather briefing and failed to notice a strong headwind. Or he fails to set the proper power setting and monitor the fuel burn rate. After hours of flying, the weather at the destination deteriorates and night falls, but the pilot decides to try to fly an approach to the airport, anyway, cutting into any fuel reserves that may have been there and leaving no extra fuel for a missed approach or a go-around or a subsequent diversion. And even though he may realize that he's low on fuel, he fails to ask for assistance from ATC and crashes short of the runway.

Not Declaring an Emergency When a Low-Fuel Situation Arises

Perhaps due to pride alone, pilots are often hesitant to declare an emergency. And when the emergency is due to nothing but poor planning, it's probably difficult for a pilot to admit to air traffic controllers that he is low on fuel. But there is no good reason not to declare an emergency in a low-fuel situation, especially if other factors are present like bad weather, an inexperienced pilot, or lack of familiarity with the surrounding area. Pilots have been known to run out of fuel trying to figure out where they are after becoming lost or disoriented and refusing to admit it and ask ATC for help.

Guessing or Assuming

It sounds like something nobody would ever do when airplanes are involved, but the number of fuel starvation accidents prove that many pilots guess at the amount of fuel in the tank before taking off, or assume that the last person who flew the airplane filled it up, or assume that because they can see fuel sloshing around in the tank somewhere down there, there is enough fuel for them to get where they're going. And some pilots guess at the fuel burn rate, thinking that they can't be that far off, but over time and distance, or with a strong headwind or a different power setting, they end up being very far off. Guessing or assuming seems like something only other people are dumb enough to do, but it happens more than you think.

Distraction

There have been aircraft accident in the past in which the pilots allowed a fuel starvation event to occur while preoccupied with something else, like fixing a landing gear problem or becoming disoriented. The adage applies here: Aviate, navigate, communicate—in that order. Troubleshooting or allowing yourself to get distracted by other people or events can lead to fixation on that particular problem or event and cause the pilot to completely disregard other important aspect of the flight—like fuel management.

Failing to Plan for Deviation From the Plan

Pilots who never plan for anything other than their one and only Plan A often find themselves in trouble as soon as Plan A goes away. Pilots should plan for the worst and hope for the best instead of just planning for the best and counting on it to work out. A pilot who never thinks that anything bad happens will not have a plan when something bad does happen. Failing to plan for deviation can result in fuel starvation if those deviations require more fuel than originally planned. A pilot's perception is often different from the reality, and assuming that everything will go according to plan which is a huge mistake.

Mechanical Problem or Failure

Very rarely, there is actually a fuel leak or a problem with the fuel system that can cause fuel starvation. In these cases, early recognition is key to recognizing and dealing with the problem. There have been aircraft accidents in the past where the pilots are too preoccupied with other things too distracted or just plain lazy, and they aren't monitoring the actual fuel burn or the status of the fuel system.

Ⅱ Words and Expressions

distinguish	*vt.* 区分
speculation	*n.* 投机;推测;思索;投机买卖
redundancy	*n.* 冗余
triple	*n.* 三倍
drain	*vt.* 排出
contaminated	*adj.* 污染的
blatant	*adj.* 明目张胆的
decimal	*adj.* 小数的;十进位的
avert	*vt.* 避免
subsequent	*adj.* 后来的,随后的
disoriented	*adj.* 无判断力的;分不清方向或目标的
slosh	*vt.* 晃动,晃荡;冲激

adage   *n.* 格言

Ⅲ   Exercises

1   Translate the following phrases into English

　(1)备降机场

　(2)燃油储备

　(3)自助泵

　(4)油量表

　(5)直率

　(6)燃油消耗率

　(7)计算

　(8)燃油泄漏

　(9)来源于

　(10)复飞

2   Translate the following sentences into Chinese

1. In the pilot training world, this means that flight instructors spend extra time on these topics and every check ride with an FAA designated examiner will certainly include at least a discussion on controlled flight into terrain and fuel management.

2. In flight training, we pay particular attention to these special emphasis areas, and we stress to students that running out of fuel happens too often by the way for anyone to get comfortable with the idea that they'll never be the one to run out of fuel.

3. Fuel mismanagement occurs when the pilot forgets to switch fuel tank when necessary, or switches to the wrong fuel tank, or just doesn't monitor the fuel burn during a flight.

4. It sounds like something nobody would ever do when airplanes are involved, but the number of fuel starvation accidents prove that many pilots guess at the amount of fuel in the tank before taking off, or assume that the last person who flew the airplane filled it up, or assume that because they can see fuel sloshing around in the tank somewhere down there, there is enough fuel for them to get where they're going.

5. Troubleshooting or allowing yourself to get distracted by other people or events can lead to fixation on that particular problem or event and cause the pilot to completely disregard other important aspect of the flight—like fuel management.

IV   Supplement Reading

Leaded Aviation Fuel and the Environment

Aircraft operating on leaded aviation gasoline (avgas) are used for many critical purposes, including business and personal travel, instructional flying, aerial surveys, agriculture, firefighting, law enforcement, medical emergencies, and express freight.

What is avgas?

Avgas is a specialized fuel used to power piston engine aircraft. Aviation gasoline is a complex mixture of relatively volatile substances known as hydrocarbons that vary widely in their physical and chemical properties. The properties of avgas must be properly balanced to give reliable and safe engine performance over an extremely wide range of aircraft operating condition. Manufacturers typically certify their engines and aircraft to run on fuels that meet American Society of Testing Materials (ASTM) Standard, or other consensus standards such as the United Kingdom's Defense Standards, or U.S. Military Standard, which govern the chemical, physical and performance properties of avgas.

The various grades of avgas are identified using the Motor Octane Number (MON) combined with the following alpha-designations to indicate lead content: low lead (LL); very low lead (VLL); or unleaded (UL).

Although there are various ASTM Standards for avgas, almost all avgas on the U. S. market today is low lead, 100 MON avgas (100LL). This grade of avgas satisfies the requirement of all piston engines using avgas, regardless of their performance level. Jet aircraft and turbine-powered, propeller aircraft do not use avgas, but instead use fuels very similar to kerosene, which does not contain a lead additive.

Why is octane so important?

Octane is a measure of the performance of a fuel as it burns in an engine combustion chamber. It is a measure of a gasoline's ability to resist detonation, or "knock". Octane is important to the safe operation of an aircraft or automobile engine. High compression, high displacement engines, such as those found in many high performance, piston engine aircraft, require high octane fuel so that detonation, which is the uncontrolled ignition of the fuel in the combustion chamber, does not damage pistons and other engine components and result in engine failure. High performance engines allow an aircraft to operate at increased speed and with more payload, but these engines require higher octane avgas. Operating aircraft or automotive piston engines on fuel with lower octane than they require may result in damage from knock, but it is generally safe to operate piston engines on fuel of a higher octane rating than their minimum requirement. In other words, it is safe to go up in octane, but not down.

What is Tetraethyl Lead (TEL)?

TEL is an organic compound that contains lead and, in small quantities, is very effective in boosting octane. The ban of TEL in automobile gas was phased in over a number of years and was largely completed by 1986 and resulted in significant reduction of lead emissions to the environment. TEL has not yet been banned for use in avgas, because no operationally safe alternative is currently available.

Is TEL Toxic?

All forms of lead are toxic if inhaled or ingested. Lead can affect human health in several ways, including effects on the nervous system, red blood cells, cardiovascular and immune system. Infants and young children are especially sensitive to even low level of lead, which may contribute to behavioral and learning problems or lower IQ. Children have increased sensitivity due to their developing nervous system.

How are aircraft emissions regulated?

Under the Clean Air Act (CAA), the EPA has the authority (in consultation with the FAA) to regulate emissions from aircraft. The CAA specifies that, in setting standards, the agencies must consider the time needed to develop required technology and cost, and not adversely impact on aircraft safety or noise. At present, there are no regulations that apply to emissions from aircraft that use leaded fuel. However, FAA enforces existing emission standards for commercial jet aircraft and engines through the certification process of engines. Commercial jet engine manufacturers have responded to requirement for emission reduction through technology change by improving jet engine design and efficiency. If the EPA finds that aircraft emissions present an endangerment to public health or welfare, they can establish limits on aircraft emission, and then the FAA has the authority to regulate aircraft emission through the development of standards for the composition and chemical or physical properties of an aircraft fuel or fuel additive.

Why keep using leaded fuel?

First and foremost, the use of leaded fuel is an operational safety issue, because without the additive TEL, the octane level would be too low for some engines and the use of a lower octane fuel than required could lead to engine failure. As a result, the additive TEL has not been banned from avgas. Aircraft manufacturers, the petroleum industry, and the FAA have worked for over a decade to find alternative fuel that meet the octane requirement of the piston engine aircraft fleet without the additive TEL. However, no operationally safe, suitable replacement for leaded fuel has yet been found to meet the need of all of the piston engine aircraft fleet.

What is FAA doing about eliminating leaded aviation fuel?

Four initiatives have been established to develop a safe unleaded replacement aviation gasoline.

First and most important, the FAA sponsored an Aviation Rulemaking Committee (ARC) involving EPA and industry stakeholders, which developed the process, cost estimate, and time line to replace existing leaded aviation fuel with unleaded solution. The final report and recommendation, known as the Unleaded Avgas Transition (UAT) Committee Final Report was published on February 17, 2012. The report is available to the public at www. faa. gov/about/ initiatives/avgas/archive. This report contains five key recommendations (fourteen additional) to facilitate the development and deployment of a replacement unleaded aviation gasoline. The plan calls for government research and development (R&D) funding and in-kind funding from industry to identify an unleaded fuel by 2018 that could be used by aircraft currently operating on leaded avgas.

Second, the FAA has established an Agency performance metric that states: "A replacement fuel for leaded aviation gasoline is available by 2018 that is usable by most general aviation aircraft." This performance metric will guide investment and decisions taken on by FAA for the coming years.

To help meet this goal, the FAA asked the world's fuel producers on June 10 to submit proposals for fuel option that would help the general aviation industry make a transition to an unleaded fuel. The FAA will assess the viability of candidate fuel in terms of their impact on the existing fleet, their production and distribution infrastructure, the environment and toxicology, and economic consideration. The FAA is asking fuel producers to submit by July 1, 2014, data packages for candidate replacement unleaded fuel formulations for evaluation by the FAA. By September. 1, 2014, the FAA will select up to 10 suppliers to participate in phase one laboratory testing at the FAA's William J. Hughes Technical Center. The FAA will select as many as two fuels from phase one for phase two engine and aircraft testing. That testing will generate standardized qualification and certification data for candidate fuel, along with property and performance data. Over the next five years, the FAA will ask fuel producers to submit 100 gallons of fuel for phase of one testing and 10,000 gallons of fuel for phase of two testing.

There are approximately 167,000 aircraft in the United States and a total of 230,000 worldwide that rely on 100 low lead avgas for safe operation. It is the only remaining transportation fuel in the United States that contains the addition of TEL.

Third, Section 910 of the 2012 FAA Modernization and Reform Act established an unleaded aviation gasoline R&D program with deliverable requirement for an R&D plan and report. The FAA has issued the Unleaded Avgas Transition (UAT) Action Plan that will integrate these three activities.

The fourth initiative involves private-sector companies that have applied for Supplemental Type Certificates for specific piston engine and aircraft model to operate with new, unleaded aviation gasoline formulation. The FAA is actively working to support all of these initiatives.

What is FAA doing in the short-term to reduce lead emission from airports?

FAA's goal for an unleaded avgas by 2018 is the long term solution that will, ultimately, allow for the elimination of lead emission from aircraft that use leaded fuel. Until such fuel can be brought to market, there are actions that FAA can coordinate with airport, aircraft owners and operators to investigate options to reduce lead emission at airport. Some of the measures that are being considered include:

1. Lower leaded fuel options It may be possible for airports to supply lower leaded fuels in current fuel distribution systems. These fuels that meet ASTM standard have been approved for being used in aircraft certified for their use and would be completely transparent in its distribution and application. Potential reductions in lead emission are as much as 19 percent since these lower level fuels have approximately 19 percent of lead content less than current fuels.

2. Consider unleaded automotive fuels as an option at airports Approximately 40 percent of piston engine aircraft are either approved or eligible to operate on automotive fuels. This unleaded fuel could represent an option for some airports, and however, any fuel used in aircraft engines must not contain ethanol; this requirement may limit the applicability of automotive fuel. This would require separate fuel systems and procedures to ensure that aircraft are fueled properly. Airport sponsors would have to make the necessary arrangement for supply, storage and distribution systems—with due consideration of the level of demand for two different fuel types—all of which may make this option be challenging both logistically and financially.

3. Safely change aircraft operations to avoid concentrated lead emission Locations for engine run-up areas could be distributed over a wider area within an airport to reduce the potential for concentrated levels of lead emission. It may also be possible to shorten taxi routes to lessen emission. Such measures would be airport-specific and have to consider operational safety as the highest priority.

4. Install vapor recovery systems Vapor recovery systems, similar to those found at automotive filling station, could be installed in bulk fuel delivery system to minimize the release of avgas vapors which contain small concentration of lead.

Unit 18

Future of Civil Aviation

I Reading Text

Your next pilot could be drone software

Would you get on a plane that didn't have a human pilot in the cockpit? Half of air travelers surveyed in 2017 said they would not, even if the ticket was cheaper. Modern pilots do such a good job that almost any air accident is big news, such as the Southwest engine disintegration on April 17. But stories of pilot drunkenness, rants, fights and distraction, however rare, are reminders that pilots are only human. Not every plane can be flown by a disaster-averting pilot, like Southwest Capt. Tammie Jo Shults or Capt. Chesley "Sully" Sullenberger.

But software could change that, equipping every plane with an extremely experienced guidance system that is always learning more. In fact, on many flights, autopilot systems have already controlled the plane for basically all of the flight. And software handles the most harrowing landings—when there is no visibility and the pilot can't see anything to even know where he or she is, but human pilots are still on hand as backups. A new generation of software pilots, developed for self-flying vehicles, or drones, will soon have logged more flying hours than what all humans ever have. By combining their enormous amounts of flight data and experience, drone-control software applications are poised to quickly become the world's most experienced pilot.

Drones that fly themselves

Drones come in many forms, from tiny quad-rotor copter toys to missile-firing winged planes, or even seven-ton aircraft that can stay aloft for 34 hours at a stretch.

When drones were first introduced, they were flown remotely by human operators. However, this merely substitutes a pilot on the ground for one aloft. And it requires significant communication bandwidth between the drone and control center, to carry real-time video from the drone and to transmit the operator's command.

Many newer drones no longer need pilots; some drones for hobbyists and photographerscan now fly themselves along human-defined routes, leaving the human free to sightsee—or control the camera to get the best view.

University researchers, businesses and military agencies are now testing larger and more capable drones that will operate autonomously. Swarms of drones can fly without needing tens or hundreds of humans to control them. And they can perform coordinated maneuvers that human controllers could never handle.

Whether flying in swarms or alone, the software that controls these drones is rapidly gaining flight experience.

Importance of pilot experience

Experience is the main qualification for pilots. Even a person who wants to fly a small plane for personal and noncommercial use needs 40 hours of flying instruction before getting a private pilot's license. Commercial airline pilots must have at least 1,000 hours before even serving as a co-pilot.

On the ground training and in flight experience prepare pilots for unusual and emergency scenarios, ideally to help save lives in situations like the "Miracle on the Hudson." But many pilots are less experienced than "Sully" Sullenberger, who saved his planeload of people with quick and creative thinking.

With software, though, every plane can have on board a pilot with as much experience—if not more. A popular software pilot system, in use in many aircraft at once, could gain more flight time each day than a single human might accumulate in a year.

As someone who studies technology policy as well as the use of artificial intelligence for drones, cars, robots and other uses, I don't lightly suggest handing over the control for those

225

additional task. But giving software pilots more control would maximize computers' advantage over humans in training, testing and reliability.

Training and testing software pilots

Unlike people, computers will follow sets of instructions in software the same way every time. That lets developers create instructions, test reactions and refine aircraft response. Testing could make it far less likely, for example, that a computer would mistake the planet Venus for an oncoming jet and throw the plane into a steep dive to avoid it.

The most significant advantage is scale: Rather than teaching thousands of individual pilots new skill, updating thousands of aircraft would require only downloading updated software.

These systems would also need to be thoroughly tested—in both real-life situations and in simulations—to handle a wide range of aviation situations and to withstand cyberattacks.

But once they're working well, software pilots are not susceptible to distraction, disorientation, fatigue or other human impairment that can create problems or cause errors even in common situation.

Rapid response and adaptation

Already, aircraft regulators are concerned that human pilots are forgetting how to fly on their own and may have trouble taking over from an autopilot in an emergency.

In the "Miracle on the Hudson" event, for example, a key factor in what happened was how long it took for the human pilots to figure out what had happened—that the plane had flown through a flock of birds, which had damaged both engines—and how to respond.

Rather than the approximately one minute it took the humans, a computer could have assessed the situation in seconds, potentially saving enough time that the plane could have landed on a runway instead of a river.

Aircraft damage can pose another particularly difficult challenge for human pilots: It can change what effects the control have on its flight.

In cases where damage renders a plane uncontrollable, the result is often tragedy. A sufficiently advanced automated system could make minute changes to the aircraft's steering and use its sensors to quickly evaluate the effect of those movements—essentially learning how to fly all over again with a damaged plane.

Boosting public confidence

The biggest barrier to fully automated flight is psychological, not technical. Many people may not want to trust their lives to computer systems. But they might come around when reassured that the software pilot has tens, hundreds or thousands more hours of flight experience than any human pilot.

Other autonomous technologies, too, are progressing despite public concern. Regulators and lawmakers are allowing self-driving cars on the roads in many states. But more than half of Americans don't want to ride in one, largely because they don't trust the technology.

And only 17% of travelers around the world are willing to board a plane without a pilot. However, as more people experience self-driving cars on the road and have drones deliver them packages, it is likely that software pilots will gain in acceptance.

The airline industry will certainly be pushing people to trust the new system: Automating pilots could save tens of billions of dollars a year. And the current pilot shortage means software pilots may be the key to having any airline service to smaller destinations.

Both Boeing and Airbus have made significant investment in automated flight technology, which would remove or reduce the need for human pilots. Boeing has actually bought a drone manufacturer and is looking to add software pilot capabilities to the next generation of its passenger aircraft (Other tests have tried to retrofit existing aircraft with robotic pilots).

One way to help regular passengers become comfortable with software pilots—while also helping to both train and test the systems—could be to introduce them as co-pilots working alongside human pilots.

Planes would be operated by software from gate to gate, with the pilots instructed to touch the control only if the system fails.

Eventually pilots could be removed from the aircraft altogether, just like they eventually were from the driverless trains that we routinely ride in airports around the world.

II   Words and Expressions

disintegration	*n.*	解体
rant	*vi.*	咆哮
drone	*n.*	无人驾驶飞机
poise	*vi.*	准备
stretch	*vt.*	拉伸
bandwidth	*n.*	带宽
swarm	*n.*	成群
steep	*adj.*	陡峭的
cyberattack	*n.*	网络攻击
disorientation	*n.*	定向障碍
steer	*vt.*	操舵
retrofit	*vt.*	改造

III   Exercises

1   Translate the following phrases into English

（1）实时视频

（2）有翼机

（3）协调动作

（4）应急场景

（5）移交

（6）飞机的转向

（7）抵御网络攻击

（8）无人驾驶汽车

（9）自动飞行

（10）无人驾驶列车

2　Translate the following sentences into Chinese

1. But stories of pilot drunkenness, rants, fights and distraction, however rare, are reminders that pilots are only human.

2. Drones come in many forms, from tiny quad-rotor copter toys to missile-firing winged planes, or even seven-ton aircraft that can stay aloft for 34 hours at a stretch.

3. On the ground training and in flight experience prepare pilots for unusual and emergency scenarios, ideally to help save lives in situations like the "Miracle on the Hudson."

4. Testing could make it far less likely, for example, that a computer would mistake the planet Venus for an oncoming jet and throw the plane into a steep dive to avoid it.

5. A sufficiently advanced automated system could make minute changes to the aircraft's steering and use its sensors to quickly evaluate the effect of those movements—essentially learning how to fly all over again with a damaged plane.

IV　Supplement Reading

Let's Go Bats

Bats have a problem: how to find their way around in the dark. They hunt at night, and cannot use light to help them find prey and avoid obstacles. You might say that this is a problem of their own making, one that they could avoid simply by changing their habit and hunting by day. But the daytime economy is already heavily exploited by other creatures such as birds. Given that there is a living to be made at night, and given that alternative daytime trades are thoroughly occupied, natural selection has favoured bats that make a go of the night-hunting trade. It is probable that the nocturnal trades go way back in the ancestry of all mammals. In the time when dinosaurs dominated the daytime economy, our mammalian ancestors probably only managed to survive at all because they found ways of scraping a living at night. Only after the mysterious mass extinction of the dinosaurs about 65 million years ago were our ancestors able to emerge into the daylight in any substantial numbers.

Bats have an engineering problem: how to find their way and find their prey in the absence of light. Bats are not the only creatures to face this difficulty today. Obviously the night-flying insects that they prey on must find their way about somehow. Deep-sea fish and whales have little or no light by day or by night. Fish and dolphins that live in extremely muddy water cannot see because, although there is light, it is obstructed and scattered by the dirt in the water. Plenty of other modern animals make their living in condition where seeing is difficult or impossible.

Given the questions of how to manoeuvre in the dark, what solutions might an engineer consider? The first one that might occur to him is to manufacture light, to use a lantern or a searchlight. Fireflies and some fish (usually with the help of bacteria) have the power to manufacture their own light, but the process seems to consume a large amount of energy. Fireflies use their light for attracting mates. This doesn't require a prohibitive amount of energy: a male's tiny pinprick of light can be seen by a female from some distance on dark night, since her eyes are exposed directly to the light source itself. However, using light to find one's own way around requires vastly more energy, since the eyes have to detect the tiny fraction of the light that bounces off each part of the scene. The light source must therefore be immensely brighter if it is to be used as a headlight to illuminate the path, than if it is to be used as a signal to others. In any event, whether or not the reason is the energy expense, it seems to be the case that, with the possible exception of some weird deep-sea fish, no animals apart from man uses manufactured light to find its way about.

What else might the engineer think of? Well, blind humans sometimes seem to have an uncanny sense of obstacles in their path. It has been given the name "facial vision", because blind people have reported that it feels a bit like the sense of touch, on the face. One report tells of a totally blind boy who could ride his tricycle at good speed round the block near his home, using facial vision. Experiments showed that, in fact, facial vision is nothing to do with touch or the front of the face, although the sensation may be referred to the front of the face, like the referred pain in a phantom limb. The sensation of facial vision, it turns out, really goes in through the ears. Blind people, without even being aware of the fact, are actually using echoes of their own footsteps and of other sounds, to sense the presence of obstacles. Before this was discovered, engineers had already built instruments to exploit the principle, for example to measure the depth of the sea under a ship. After this technique had been invented, it was only a matter of time before weapons designers adapted it for the detection of submarines. Both sides in the Second World War relied heavily on these devices, under such codenames as Asdic (British) and Sonar (American), as well as Radar (American) or RDF (British), which uses radio echoes rather than sound echoes.

The Sonar and Radar pioneers didn't know it them, but all the world knows that bats, or rather natural selection working on bats, had perfected the system tens of millions of years earlier, and their "radar" achieves feats of detection and navigation that would strike an engineer dumb with admiration. It is technically incorrect to talk about bat "radar", since they do not use radio waves. It is sonar. But the understanding mathematical theories of radar and sonar are very similar, and much of our scientific understanding of the details of what bats are doing has come from applying radar theory to them. The American zoologist Donald Griffin, who was largely responsible for the discovery of sonar in bats, coined the term "echolocation" to cover both sonar and radar, whether used by animals or by human instruments.

Fireflies use their light for attracting mates. This doesn't require a prohibitive amount of energy: a male's tiny pinprick of light can be seen by a female from some distance on dark night, since her eyes are exposed directly to the light source itself. However, using light to find one's own way around requires vastly more energy, since the eyes have to detect the tiny fraction of the light that bounces off each part of the scene. The light source must therefore be immensely brighter if it is to be used as a headlight to illuminate the path, than if it is to be used as a signal to others. In any event, whether or not the reason is the energy expense, it seems to be the case that, with the possible exception of some weird deep-sea fish, no animals apart from man uses manufactured light to find its way about.

KEYS FOR EXERCISES

UNIT 1

1. Translate the following phrases into English

(1) fixed wing aircraft

(2) surface – to – air missiles

(3) unmanned aerial vehicle

(4) point – to – point flight

(5) Jet Age

(6) power – to – weight ratio

(7) military use

(8) heavier – than – air flight

(9) long – range bombers

(10) Third law of motion

2. Translate the following sentences into Chinese

(1) 在18世纪晚期,乘热气球在欧洲成为一种主要的"时尚",人们首次详细了解了高度和大气之间的关系。

(2) 实施科学的空气动力学实验,用于证实阻力和流线型,压力中心的移动,以及在弯曲机翼表面上升力增加这些现象。

（3）飞机从由木头和塑料制成的低功率双翼飞机发展到由铝制成的光滑、高功率的单翼机,是基于第一次世界大战期间雨果·强克斯(Hugo Junkers)的创新工作,以及美国设计师威廉·布什内尔·斯图特(William Bushnell Stoutout)和苏联设计师安德烈·图波列夫(Andrei Tupolev)的研究成果的应用。

（4）即使是一支中等规模的远程轰炸机编队,也能对敌人造成致命的打击,因此人们做出了很大的努力来制定对抗措施。

（5）现代航空飞机结构由固定机翼、机身和尾部组件组成。

UNIT 2

1. Translate the following phrases into English

（1）cut – out

（2）fuel tank

（3）cross-section

（4）shear force

（5）bending moment

（6）keel beam

（7）monocoque structure

（8）three – dimensional space

（9）center of gravity

（10）bounce up

2. Translate the following sentences into Chinese

（1）一种很好的使用金属薄板表面的方法是做成一个薄壁圆柱体,称为单壳结构。

（2）挤压加筋板是通过挤压一定形状的有开口的受热和粘性的材料制成的。

（3）现在钢丝绳穿过在肋骨上凿出的孔,这也不可避免地造成了结构的削弱。

（4）由于飞机的前部和后部的重量,在中心部分产生了较大的弯矩。

（5）在大多数飞机中,尾部单元的唯一功能是提供所需的稳定性和控制能力。

UNIT 3

1. Translate the following phrases into English

(1) radio communication

(2) put into effect

(3) controlled airspace

(4) Visual Flight Rules

(5) in the vicinity of

(6) in compliance with

(7) glass – enclosed

(8) tower cab

(9) onactive runways

(10) inbound aircraft

2. Translate the following sentences into Chinese

(1) 早在20世纪20年代,最早的空中交通管制员就在机场附近用灯光和旗帜对飞机进行人工导航,而信标台和闪光灯则被放置在纵横航线上,目的是建立最早的航线。

(2) 直到美国联邦航空局成立后,才对美国领空进行了全面监管,这是很偶然的,因为喷气式发动机的出现,突然导致了大量高速飞机的产生,降低了飞行员的误差允许范围,实际上要求制定一些规则,让每个人都能在空中保持良好的距离和操作安全。

(3) 这样,享受娱乐飞行的飞行员想要飞行一段时间并且不受美国联邦航空局的限制,只能停留在在非管制空域,即低于365 m,而想要空中交通管制提供保护的飞行员则可以轻松进入管制空域。

(4) 如果空中交通管制中心(ARTCC)将间隔责任委托给空中交通管制塔台在运作上能更具有优势时,塔台和管制中心的代表将共同起草一份适当的协议书。

(5) 地面管制员在塔台玻璃幕墙的部分即塔台工作室工作,负责将为在坡道、滑行道和任何非活动跑道上运行的飞机和车辆建立间隔。

UNIT 4

1. Translate the following phrases into English

(1) Maximum Takeoff Weight

(2) structural damage

(3) Maximum Brake Release Weight

(4) structural limitation

(5) Maximum Landing Weight

(6) reference datum

(7) Runway gradient

(8) maximum allowable weight for takeoff

(9) Handbook of Aeronautical Knowledge

(10) climb performance

2. Translate the following sentences into Chinese

(1) 质量非常大的飞机不会有很好的爬升率,因此计算飞机在特定起飞重力下的爬升梯度和爬升速率尤为重要。

(2) 飞机在高海拔地区的性能会下降,这意味着全功率是不太可能实现。

(3) 飞机的空载重力是飞机本身的重力,不包括乘客、行李或燃料。

(4) 需要注意的是,仅仅因为一架飞机获得了特定最大起飞重力的认证,其实并不意味着飞机总是可以在这个最大起飞重力下起飞。

(5) 密度高度越高,即修正非标准温度后的压力高度越高,其飞机性能越差。

UNIT 5

1. Translate the following phrases into English

(1) International Air Transport Association

(2) Global Business Travel Association

(3) Yangtze Delta region

(4) flight delay and cancellation

(5) airspace constraint issue

（6）low – altitude airspace

（7）hang around in airport

（8）domestic airline industry

（9）opening the emergency doors

（10）high – speed train network

2. Translate the following sentences into Chinese

（1）阿姆斯特朗的沮丧情绪在中国的商务旅行者中很常见,这是因为中国长期以来,且至少在有充分记录的时间内,存在过度拥挤的航行空域,导致航空旅行在近年来变得越来越混乱和不可靠。

（2）这不仅给客机造成了每日的瓶颈,而且当军方想要进行演习,它还会偶尔导致大量航班取消。例如在 2014 年的夏天,有 12 个机场,其中包括两个在上海,因为期三周的军事训练而被命令飞行交通流量减少了 25%。

（3）中国通过增加每小时可以安全起飞和着陆飞机的数量,也正积极建设更多的机场和努力提高其现有设施的效率,在中国航空领域工作 30 年的,华盛顿国家公务航空协会的首席运营官史蒂文·布朗(Steven Brown)这样说。

（4）阿姆斯特朗说,他过去被迫为航班取消或延误找借口,他说这是"非常可怕的",尤其是当他在北京会见银行监管机构人员时。

（5）他说,为了实现这一目标,中国在航空业正在采用最佳的操作方法,从训练提高机组人员在操纵飞机时更有效率,到提高空中交通管制能力和跑道灯光导引能力。

UNIT 6

1. Translate the following phrases into English

（1）traffic volume

（2）runway capacity

（3）tactical decision – making

（4）key performance area

（5）Operational Efficiency

（6）Information Exchange

（7）Collaborative Decision – Making

（8）Minimizing taxi queues

(9)Aviation System Block Upgrades

(10)ground handling

2. Translate the following sentences into Chinese

(1)在财政和环境的限制下,需要在现有预算和现有基础设施的范围内尽量扩大资产。

(2)债权人是从他们的资产中获得最大的收益和服务,同时改进的容量和成本效益。

(3)如果在所有利益相关者之间没有共同的情景意识,以及缺乏在战术和战略上进行适当计划的能力,延迟可能会加剧和升级,并进一步恶化局势,形成一种在利益相关者之间相互指责的文化。

(4)如果由于登机口电脑故障,航班将比原计划晚 15 分钟登机,该信息将同时提供给匝道管理、登机口分配、拖车调度员、航路和终端区 ATCO,以及所有可能处在目的地机场航班受到影响情况下的团体。

(5)这将使计算新的最佳开车时序时,考虑到抵达的飞机、个体滑行时间、除冰要求和周围空域的容量限制。

UNIT 7

1. Translate the following phrases into English

(1) bird strike

(2) emergency response

(3) airport emergency plan (AEP)

(4) noise complaints

(5) emissions testing

(6) Hazardous weather operations

(7) public safety

(8) terrorist attack

(9) advisory circular?

(10) plane crash

2. Translate the following sentences into Chinese

(1)运营经理熟悉所有进出的空中交通量,乘客数量和燃料使用情况。他们通常负责实施规章制度,确保安全手册和程序是最新的,在必要情况下准备以及实施计划。

（2）典型的机场应急计划包括几个不同的组成部分,通常由机场经理或应急响应协调员创建和实施。

（3）一旦确定了潜在的危险,并完成了风险评估,机场应急响应协调员就可以开始制定针对特定场景的计划。例如对于飞机坠毁,会有不同于炸弹威胁的应急计划。

（4）帮助管理人员和协调人员制定最佳计划的方法之一是反复实践该计划,详尽探讨不同的场景,并利用所有可用的资源,确保在紧急情况发生时所有部门都知道自己的角色。

（5）在这些部门的副总裁之下,都是一些较小部门的副总裁,比如 IT、人力资源、环境事务、公共安全、公共事务、市场营销、特许经营、停车场等。

UNIT 8

1. Translate the following phrases into English

（1）deduced reckoning

（2）crosswind correction angle

（3）Low Earth Orbit

（4）ground track

（5）true heading

（6）magnetic north

（7）magnetic heading

（8）magnetic compass

（9）automatic direction finder

（10）compass deviation

2. Translate the following sentences into Chinese

（1）当飞机经过这些地标时,飞行员会记录任何偏离计划飞行路线的情况,并调整飞机航向,使其返回预定航线。

（2）在空速变化或飞机转弯时,磁罗盘会指示错误,这些特殊的误差被称为加速度和转向误差。

（3）此外,放置在指南针附近的金属或磁化物体(如手电筒、剪贴板或螺丝刀)将会改变局部磁场,引起指南针误差。

（4）随着飞机电子技术的进步,手动操作的 NDB 接收器很快就被自动测向仪(ADF)所取代,ADF 可以通过电子方式确定 NDB 的方位,并将这些信息显示给飞行员。

（5）当地面的 DME 应答器接收到这个脉冲时,它触发一个以不同频率传输的编码应答。当询问器接收到这个脉冲时,经过的范围时间是通过电子计算的。

UNIT 9

1. Translate the following phrases into English

（1）visual flight rules

（2）instrument meteorological conditions

（3）carbon emission

（4）Wide Area Augmentation System

（5）restricted visibility

（6）situational awareness

（7）Traffic Information Service – Broadcast

（8）flight route

（9）controlled to an acceptable level

（10）Avionics malfunctions

2. Translate the following sentences into Chinese

（1）它将使飞机能够飞行更短的航线,缓解拥堵,减少碳排放,并为飞机运营商节省时间和金钱。

（2）ADS－B 的工作原理是利用卫星信号和飞机航空电子系统来解析飞机数据,并将其连续地、实时地以广播的方式发送给空中交通管制员。

（3）Capstone 项目确定,在使用 ADS－B 时,过多的机头向下时间可能会导致频繁的态势感知丧失,尽管在这种情况下事故可能很少见,但由此产生的事故可能是灾难性的。

（4）飞行员必须通过培训和意识来承担尽可能多地减少这种风险的责任。

（5）ADS－B 一旦投入使用,就会成为一个有价值的工具,为我们的空中交通管制员和飞行员提供最迄今为止准确的数据。如果在全国范围内实施,好处是积极的。

UNIT 10

1. Translate the following phrases into English

（1）Controlled airspace

（2）Special-use airspace

（3）Federal Aviation Administration

（4）flight service stations

（5）Air Route Traffic Control Center

（6）Terminal Radar Approach Control

（7）Mean Sea Level

（8）Automatic Dependent Surveillance-Broadcast

（9）Global Positioning System

（10）UHF（ultra-high frequency）

2. Translate the following sentences into Chinese

（1）除了已经使用多年的多种不同技术外,航空工业还在不断开发新技术,使该系统对飞行员和管制员来说更有效、更容易和更安全。

（2）国家空域系统(NAS)是在商业航空发展初期建立的,目的是使飞机安全有效地从A地飞到B地。

（3）NextGen项目旨在通过寻找处理增加的交通流量和改进整个系统的方法来提高当前的NAS系统。

（4）然而,由于气象站点是在地面上(不随飞机移动),因此该站点报告的压力可能与飞机实际位置的压力不同,从而影响高度计读数的准确性。

（5）密度高度对于决定飞机的性能,或飞机在特定条件下怎样飞行非常重要。

UNIT 11

1. Translate the following phrases into English

（1）Air Defense Identification Zone

（2）Military Operations Area

（3）Standard Instrument Departure

（4）Flight levels

（5）Standard Terminal Arrival Route

（6）Mean Sea Level

（7）transition altitude

（8）divert to alternate

（9）minimum weather conditions

（10）step climb

2. Translate the following sentences into Chinese

（1）规划过程必须谨慎,只包括能够达到预期燃料负荷和飞机总质量,以及有能力处理正在飞行的飞机类型的备降机场。

（2）储备燃油可以作为在目的地飞机上剩余的额外燃油来计划,也可以假设是在飞行过程中燃烧的(可能是由于实际飞机和飞行性能数据之间的差异而没有考虑到得)。

（3）航空公司有一组相关的标准化飞行等级(有时称为"飞行模型"),在飞行途中必须使用这些等级。

（4）建议国内航班最少在起飞前1个小时递交飞行计划,对于国际航班要在起飞前3小时前递交飞行计划。

（5）计划员使用预测的天气和飞机重量作为飞行性能数据的输入,以估计到达目的地所需的燃油。

UNIT 12

1. Translate the following phrases into English

（1）categories of airspace

（2）lack of terrain definition

（3）Marginal VFR conditions

（4）Aeronautical Information Manual

（5）go around

（6）Federal Aviation Regulations

（7）loss of control

（8）weather patterns

（9）turn around

（10）visual references

2. Translate the following sentences into Chinese

（1）事实上,除了对违反 FAA 规定的担忧,飞行员真正应该关心的是,他们是否有足够的视觉参考资料,让飞机保持右侧飞行,并进行视觉导航。

（2）很少有飞行员能在阿拉斯加北部享受冬季飞行的乐趣,还有那些能很快学会在边缘条件下要么留在地面要么联系仪表飞行技能从 A 地飞到 B 地的人。

（3）在这些情况下,可能最重要的困难是飞行员要认识到他或她正在进入仪表气象条件,即使能见度和上限满足目视飞行规则的基本要求。

（4）如果你爬上一架飞机,然后起飞,告诉自己你可以随时掉头,虽然折叠卡片和做 180 度转弯更难,但这种关键的机动必须始终作为一种选择。

（5）在计划飞行时,不仅要考虑可能遇到的天气条件,还要考虑沿途可能遇到的地形和任何障碍,以及在这些条件下能否保持目视飞行。

UNIT 13

1. Translate the following phrases into English

（1）human – machine system

（2）flight information region

（3）ATC responsibilities

（4）controller's workload

（5）expression of speech

（6）initial training of new controllers

（7）retraining of qualified controllers

（8）quantitative and qualitative information

（9）shift work

（10）Individual differences

2. Translate the following sentences into Chinese

（1）未来的空中交通管制系统将试图通过更简洁和处理过的信息来帮助管制员,提醒管制员注意异常和冲突,并通过提供航路冲突预测解决方案来帮助预测未来的问题。

（2）在管理标准化测试之前,首先要确定管制员中特定的理想属性,以便对候选对象进行相应的评估。

（3）当 ATC 发生变化以及发现新的可检测的人体维度时,这些选择过程一定要随着不断进化,从而使该过程不会变得无关紧要。

（4）因此,培训应遵循建议的人为因素程序和做法,以满足他们的个人需要,并使他们了解其局限性和能力,同时能够选择适当的工具来提高他们的工作绩效。

(5)培训必须鼓励不断扫视和预警技术,这样管制员才能不受制于视野狭窄,过于专注于特定的任务和忽视其他重要的任务,并且必须使管制员对两种类型的灯光和交通拥挤做好准备,这样他才能够自信而又安全的处理每一种情况。

UNIT 14

1. Translate the following phrases into English

(1) risk factor associated with flying

(2) risk management and safety awareness

(3) medical certificate

(4) flight training standards

(5) regulations and company policies

(6) an emotionally stable state of mind

(7) decreases reaction time

(8) mental or physical impairment

(9) financial troubles

(10) inadequate oxygen levels

2. Translate the following sentences into Chinese

(1)它来自疲劳、剧烈运动、变形或更改时区等。不健康的饮食习惯、疾病和其他身体疾病也属于这一类。

(2)然而飞行员应该铭记,他们可以遵循"从酒瓶子到油门8小时"的规则,但这样仍然不适合飞行。宿醉在驾驶舱内也很危险,其影响类似于醉酒或生病:恶心、呕吐、极度疲劳、注意力不集中、头晕等。

(3)情绪在大多数时候是可以被抑制和控制的,但它们也可以很容易地重新浮现,尤其是在面对压力的情况下。

(4)疲劳的影响是累积的,这意味着随着时间的推移,轻微的睡眠剥夺对飞行员来说也是危险的。

(5)飞行员在管理疲劳时还应考虑时间变化、时差和昼夜排班选项。

UNIT 15

1. Translate the following phrases into English

(1) first officer

(2) simulator training

(3) marriage virtues

(4) long-term education process

(5) mentally fatiguing

(6) back up

(7) towing banner

(8) Information overload

(9) check out

(10) leadership position

2. Translate the following sentences into Chinese

(1)飞行员只需要传达一个事实:夏威夷的天气很好并且他们确实在酒店的酒吧里享受了一杯鸡尾酒;他们很累并且仍然忠于他们的婚姻。

(2)训练结束后,飞行员回家的时间通常很短,然后他们就会去他们的备用地点,这意味着他们必须住在机场附近,以防被突然被叫去飞行。

(3)对许多飞行员来说,通勤是他们工作的一部分,他们甚至在计划开始前就飞到指定的住所,并在私人时间内完成,通常会在计划的飞行行程开始前增加一天的时间。

(4)当飞行员回到家的时候,他或她可能会不想离开家,这就是为什么飞行员可能会拒绝在他或她休周假时来一次家庭度假的想法。

(5)世上有很多婚姻幸福的飞行员,但幸福婚姻的秘诀与其说是飞行员的日程安排,不如说是基本的婚姻美德。了解飞行员的生活方式仅仅是个开始。

UNIT 16

1. Translate the following phrases into English

(1) Meteorological Services in the Terminal Area

(2) Commission for Aeronautical Meteorology

(3) flight route

(4) approach area

(5) Air Traffic Meteorology Center

(6) Aerodrome Meteorological Observation and Forecast Study Group

(7) Trend – type Landing Forecast

(8) Aerodrome Forecast

(9) convective weather

(10) low ceiling/visibility weather

2. Translate the following sentences into Chinese

(1) 近年来,随着航空交通量的增加,机场及航路容量的限制,以及气象科学(例如数值天气预报及"临近预报"技术)的发展,为支援航空交通管理而设计的气象产品不断涌现。

(2) 这种天气数据的编码方面在 20 世纪中期是必要的,主要是为了克服遗留电信系统中严重的带宽限制。

(3) 自那以来,当气象学家试图将现有天气信息的具体细节传达给现代飞机运营商时,它已成为气象学家们的一个严重制约因素。

(4) 为了解决这些新的和不断发展的 ATM 用户需求,并避免昂贵的并行开发相似的的数据格式各不相同,令人困惑的气象产品,气象组织已成立了一个专家小组,以便与民航组织密切合作,制定适当的 MSTA 建议。

(5) 它将为支持代码分隔多路通讯制的不同用户组提供数据共享以及态势感知,并可通过适当的文本描述加以补充,并简化/压缩,以方便飞机驾驶舱的上行链路传输。

UNIT 17

1. Translate the following phrases into English

(1) alternate airports

(2) fuel reserves

(3) self-serve pumps

(4) fuel gauges

(5) flat-out

(6) fuel burn rate

(7) figure out

（8）fuel leak

（9）stems from

（10）go-around

2. Translate the following sentences into Chinese

（1）在飞行员培训领域,这意味着飞行指导员要在这些主题上花费额外的时间,而且FAA 指定的审查员每次检查飞行时,肯定会至少包括一次有关地形和燃油管理的管制飞行的讨论。

（2）在飞行训练中,我们特别关注这些重点的领域,我们向学生强调,燃油耗尽的情况发生得太频繁了,以至于任何人都不会认可这样的想法:他们永远不会遇到燃油耗尽的情况。

（3）如果飞行员在必要的时候忘记切换油箱,或者切换到错误的油箱,或者只是在飞行中没有监控燃油燃烧时,就会发生燃油管理不当。

（4）听起来当涉及飞机时没人会这样做,但是一系列燃油枯竭的事故证明了许多飞行员在起飞前只是猜测油箱中的油量,或是假设最后一个开过飞机的人已经加满了油箱,或是他们只是看到在油箱底部有燃油在晃动,就认为有足够的燃油来抵达目的地。

（5）故障排除或让自己被其他人或事件分心会导致你专注于某个特定的问题或事件,并可能导致飞行员完全忽视飞行中的其他重要方面——比如燃油管理。

UNIT 18

1. Translate the following phrases into English

（1）real-time video

（2）winged planes

（3）coordinated maneuvers

（4）emergency scenarios

（5）hand over

（6）aircraft's steering

（7）withstand cyberattacks

（8）self-driving cars

（9）automated flight

（10）driverless trains

2. Translate the following sentences into Chinese

（1）但是飞行员酗酒、咆哮、打架和分心的故事，无论多么罕见，都提醒我们飞行员只是人。

（2）无人机有多种形式，从小型四旋翼直升机玩具到发射导弹的带翼飞机，甚至是能在空中连续飞行 34 个小时的 7 吨重的飞机。

（3）地面训练和飞行经验为飞行员应对不同寻常的紧急情况做好了准备，在最理想的情况下挽救生命，就像"哈德逊奇迹"这样。

（4）例如，通过测试可以使计算机将金星误认为迎面而来的喷气机，使飞机进入一个陡峭的俯冲以避开它。

（5）一个足够先进的自动化系统可以对飞机的转向做出微小的改变，并利用其传感器快速评估这些动作的效果——重要的是学习如何在受损的飞机上再次飞行。